T0338508

A SYSTOLIC ARRAY
OPTIMIZING COMPILER

THE KLUWER INTERNATIONAL SERIES
IN ENGINEERING AND COMPUTER SCIENCE

VLSI, COMPUTER ARCHITECTURE AND
DIGITAL SIGNAL PROCESSING

Consulting Editor

Jonathan Allen

A SYSTOLIC ARRAY
OPTIMIZING COMPILER

by

Monica S. Lam
Carnegie Mellon University

with a foreword by

H.T. Kung

KLUWER ACADEMIC PUBLISHERS
Boston/Dordrecht/London

Distributors for North America:
Kluwer Academic Publishers
101 Philip Drive
Assinippi Park
Norwell, Massachusetts 02061, USA

Distributors for the UK and Ireland:
Kluwer Academic Publishers
Falcon House, Queen Square
Lancaster LA1 1RN, UNITED KINGDOM

Distributors for all other countries:
Kluwer Academic Publishers Group
Distribution Centre
Post Office Box 322
3300 AH Dordrecht, THE NETHERLANDS

Library of Congress Cataloging-in-Publication Data

Lam, Monica S.
  A systolic array optimizing compiler.

  (Kluwer international series in engineering and
computer science.   VLSI, computer architecture, and
digital signal processing)
  Includes index.
  1. Compilers (Computer programs) 2. Systolic
array circuits. I. Title. II. Series.
QA76.76.C65L36   1989   005.4′53   88-13904
ISBN 0-89838-300-5

To my sister Angie

Table of Contents

List of Figures

List of Tables

Foreword by H. T. Kung

Generally speaking, computer designers use two methods to build high performance computers. One is to use fast hardware components. But this approach has the drawback that high speed parts may be expensive and sometimes may even have to be custom made. The second method, which avoids this drawback, is to use parallel architectures. That is, the hardware provides multiple functional units that can operate simultaneously. As today's high performance machines are already approaching the speed limit of hardware components, more and more designs will have to rely on parallel architectures for the next level of speed up.

To use a machine based on a parallel architecture, we need to schedule the multiple functional units in the machine to implement desired computations. This scheduling task is one of the well known difficult problems in computer science. Monica Lam in this monograph describes a compiler that shield most of the difficulty from the user, and can generate very efficient code.

These results are based on Monica Lam's implementation of an optimizing compiler on the Warp systolic computer at Carnegie Mellon University. She has extended the state-of-the-art compiler techniques such as software pipelining, and devised several new optimizations. These are clearly described and carefully evaluated in the monograph.

I hope that more books of this nature, that are based on real experiences on real computer systems, can be published. These books are especially appreciated in graduate courses in computer systems. I compliment Kluwer's efforts in this area.

H. T. Kung

Preface

This book is a revision of my Ph.D. thesis dissertation submitted to Carnegie Mellon University in 1987. It documents the research and results of the compiler technology developed for the Warp machine.

Warp is a systolic array built out of custom, high-performance processors, each of which can execute up to 10 million floating-point operations per second (10 MFLOPS). Under the direction of H. T. Kung, the Warp machine matured from an academic, experimental prototype to a commercial product of General Electric. The Warp machine demonstrated that the scalable architecture of *high-performance*, *programmable* systolic arrays represents a practical, cost-effective solution to the present and future computation-intensive applications. The success of Warp led to the follow-on iWarp project, a joint project with Intel, to develop a single-chip 20 MFLOPS processor. The availability of the highly integrated iWarp processor will have a significant impact on parallel computing.

One of the major challenges in the development of Warp was to build an optimizing compiler for the machine. First, the processors in the array cooperate at a fine granularity of parallelism, interaction between processors must be considered in the generation of code for individual processors. Second, the individual processors themselves derive their performance from a VLIW (Very Long Instruction Word) instruction set and a high degree of internal pipelining and parallelism. The compiler contains optimizations pertaining to the array level of parallelism, as well as optimizations for the individual VLIW processors. The compiler has been used extensively within and outside the Carnegie Mellon University research community, and the quality of the code produced has been surprisingly good.

This research confirms that compilers play a major role in the development of novel high-performance architectures. The Warp compiler was developed concurrently with the Warp architecture. It provided essential input to the architecture design that made Warp usable and effective.

Acknowledgments

I want to thank H. T. Kung for the past years of advice; I am fortunate to have him as an advisor. He has taught me much about many different aspects of research, and has created an exciting project and a productive research environment. His guidance and enthusiasm helped make my thesis work both rewarding and enjoyable.

I would also like to express my thanks to Thomas Gross. He has put in tremendous efforts into the compiler, in both the technical and managerial aspects. Together with the rest of the compiler team, C. H. Chang, Robert Cohn, Peter Lieu, Abu Noaman, James Reinders, Peter Steenkiste, P. S. Tseng, and David Yam, he transformed my code generator from an academic experiment into a real-life system.

All members of the Warp project have contributed to this research. Thanks go to the hardware team for making a machine that runs reliably. In particular, Onat Menzilcioglu and Emmanuel Arnould had done a tremendous job designing and building the Warp cell and the interface unit. I wish to thank Marco Annaratone for his dedication in his work on the external host, and Bernd Bruegge for creating a fine programming environment. I have to thank everybody in the project for using and putting up with the W2 compiler. Jon Webb is one of our primary customers; if not for him, I would not have the 72 programs for evaluating the compiler.

I want to thank Mosur Ravishankar, who has painstakingly read through drafts of my thesis, and has given me plenty of useful comments. I'd like to thank my colleagues and friends for a wonderful time at Carnegie Mellon University. Last, but not least, my thanks go to my parents and family for their support and love.

The research was supported in part by Defense Advanced Research Projects Agency (DOD) monitored by the Space and Naval Warfare Systems Command under Contract N00039-85-C-0134, and in part by the Office of Naval Research under Contracts N00014-80-C-0236, NR 048-659, and N00014-85-K-0152, NR SDRJ-007.

1
Introduction

Since Kung and Leiserson introduced "systolic arrays" in 1978, tremendous amounts of research effort have gone into systolic algorithm research. Over the last decade, researchers have developed many systolic algorithms in many important numerical processing areas, including signal processing and scientific computing, demonstrating that systolic architecture is highly suited to deliver the bandwidth required of those applications. The original concept of systolic computing is to map systolic algorithms directly onto custom hardware implementations, and the cost of implementation has limited the practice of systolic computing to a few instances of dedicated systolic hardware. Most systolic algorithms remained as paper designs.

The CMU's Programmable Systolic Chip (PSC) project was the first to address some of the lower level architecture and implementation issues of programmable systolic arrays, and proposed a building block for a set of systolic algorithms. This work showed that there was a wide gap between the theoretical systolic array machine model and practical, programmable systems and exposed many practical issues that must be addressed before systolic computing could become a reality. For example, floating-point arithmetic capability is absolutely necessary for systolic arrays to be employed in signal processing or scientific computing. The cost of hardware logic to implement floating-point arithmetic makes programmability even more attractive.

The introduction of single-chip floating-point processors at the end of 1983 sparked off the design of the Warp machine here at Carnegie Mellon. The Warp machine is a high-performance, programmable systolic array. With each processor capable of a peak computation rate of 10 million floating-point operations per second (10 MFLOPS), a 10-cell array can deliver a peak aggregate bandwidth of 100 MFLOPS. With the assistance of our industrial partner, General Electric, the first 10-cell array was completed in early 1986. The Warp machine is the first commercial programmable systolic array.

The Warp machine is an important landmark in systolic computing research. It is an execution vehicle for the numerous systolic algorithms that have been designed and never implemented. More importantly, we have greatly extended the concept of systolic computing and showed that systolic arrays of *programmable* and *high-performance* processors can deliver execution speeds that rival existing supercomputers in the numerical processing arena.

To integrate the concepts of programmability and high-performance cells into systolic processing, we carefully scrutinized every characteristic typical of theoretical systolic array models with considerations to hardware, software and applications. As a result, the Warp machine represents a significant departure from the conventional systolic array architecture. The Warp architecture retains the high bandwidth, low latency inter-cell communication characteristic, and thus possesses the scalability property of systolic arrays. On the other hand, many of the assumed systolic array characteristics were constraints imposed by direct implementation of the algorithm in custom VLSI circuitry. As an example, to keep the logic simple and regular, all cells in the array must repeat the same operation in lock step throughout the computation. Such constraints were removed whenever feasible. This architecture study is significant in fleshing out the abstract machine model and updating the model with realistic implementation constraints.

But what is the cost of programmability? Systolic arrays are known for their efficiency; can a programmable machine be as efficient as dedicated hardware? To ensure that the machine has comparable raw processing capability, the Warp cells employ optimization techniques such as pipelining and parallel functional units, and the multiple functional units are directly accessible to the user through a VLIW (Very Long Instruction Word) instruction set. Each individual functional unit is controlled by a dedicated field in the instruction word. There is little difference between this architecture and a microprogrammed, dedicated machine. From the hardware standpoint, the architecture is programmable and the cost of additional hardware to support programmability is low.

In practice, however, a machine can be called programmable only if applications can be developed for the machine with a reasonable effort. With its wide instruction words and parallel functional units, each individually Warp cell is already difficult to program. Moreover, the multiple cells in the array must be coordinated to cooperately in a fine-grained manner in a systolic algorithm. Can we generate programs that efficiently use the parallelism at both the array and the cell level? This question is the focus of this work.

The thesis of this work is that systolic arrays of high-performance cells can be programmed effectively using a high-level language. The solution has two components: (1) a systolic array machine abstraction for the user and the supporting compiler optimizations and hardware features, and (2) code scheduling techniques for general VLIW processors.

The user sees the systolic array as an array of sequential processors communicating asynchronously. This machine abstraction allows the user to concentrate on high-level systolic algorithm design; the full programmability of the processor is accessible to the user through high-level language constructs. Optimizing compilation techniques translate such user programs into efficient code that fully exploits the machine characteristics of the high-performance processors. These techniques include systolic array specific optimizations as well as general scheduling techniques for highly pipelined and parallel processors.

1.1. Research approach

The ideas presented in this book have been implemented and validated in the context of the Warp project. The project started in 1984. The architecture of the machine has gone through three major revisions: a 2-cell prototype was constructed in June 1985, two identical 10-cell prototypes were completed by our industrial partners, General Electric and Honeywell, in the first quarter of 1986, and the first production machine arrived in April 1987. The production version of the Warp machine is currently being marketed by General Electric. Since 1986, in collaboration with Intel Corporation, we have improved the architecture, and started the design and development of a single-chip processor called iWarp. The component is expected to be available by the fourth quarter of 1989.

The research presented here consists of several components. First, we have developed a new machine abstraction for systolic arrays, and captured the abstraction by a programming language called W2. This language hides the internal parallelism and pipelining of the cells. The abstraction was designed together with the code scheduling techniques to ensure that effective code can be generated. The scheduling algorithms were implemented as part of a highly optimizing compiler supporting the language. In the course of the development of the machine abstraction and compiler, necessary hardware support has been identified, some of which was incorporated in the prototypes and some in the production machine. Finally, a large set of performance data on the machine and the compiler has been gathered and analyzed.

The compiler has been functional since the beginning of 1986. It has been used extensively in many applications such as low-level vision for robot vehicle navigation, image and signal processing, and scientific computing [3, 4]. An average computation rate of 28 MFLOPS is achieved for a sample of 72 application programs. They are all programs with unidirectional data flow that achieve a linear speed up with respect to the number of cells used. Analysis of the programs shows that, on average, the scheduling techniques speed up the programs by about a factor of three.

1.2. Overview of results

The results of this research have two major components: a machine abstraction for systolic arrays, and compiler code scheduling techniques. While this research is undertaken in the context of the Warp machine, the results are applicable to other architectures. The proposed machine abstraction and array level optimizations are useful for systolic array synthesis where either the algorithm and/or machine model is too complex for existing automatic synthesis techniques. The cell level optimizations are applicable for high-performance processors whose internal pipelining and parallelism are accessible at the machine instruction level.

1.2.1. A machine abstraction for systolic arrays

This work proposes an intermediate machine abstraction in which computation on each cell is made explicit and cells communicate asynchronously. That is, a cell communicates with its neighbors by receiving or sending data to a dedicated queue; a cell is blocked when trying to send to a full queue or receive from an empty queue. The abstraction that each cell is a simple sequential processor allows the user, or a higher level tool, to concentrate on the utilization of the array. Its asynchronous communication primitives allow the user to express the interaction between cells easily.

A surprising result is that while the asynchronous communication model provides a higher level of abstraction to the user than synchronous models, it is also more amenable to compiler optimizations. The high-level semantics of the asynchronous communication model is exploited by the compiler to relax the sequencing constraints between communication operations within a cell. Representing only those constraints that must be satisfied in a manner similar to data dependency constraints within a computation, general code motions on the communication operations can be applied to minimize the execution time on each cell. This approach allows us to generate highly efficient code for systolic arrays of unidirectional data flow from complex systolic programs.

The asynchronous communication model is useful, and in fact recommended, even for hardware implementations without direct support for dynamic flow control. Many systolic array algorithms, including all previously published ones, can be implemented without this dynamic flow control. We can first optimize the program by pretending that there is dynamic flow control; compile-time flow control is then provided using an efficient algorithm.

1.2.2. Cell level optimizations

The high computation rate of the Warp processor is derived from its heavily pipelined and parallel data path and a horizontal instruction format. The data path consists of multiple pipelined functional units, each of which is directly controlled by a dedicated field in the machine instruction. Computation that can proceed in parallel in a single cycle includes seven floating-point multiplications, seven floating-point additions, eight register accesses, two memory accesses, four data queue accesses and one conditional branch operation. The potential processing capacity available in this machine organization is tremendous. The code scheduling problem for the Warp processors has been studied in the context of VLIW (very long instruction word) architectures [20], and trace scheduling has been the recommended approach [19]. This work suggests an alternative approach to scheduling VLIW processors: *software pipelining* and *hierarchical reduction*.

Software pipelining is a scheduling technique that exploits the repetitive nature of innermost loops to generate highly efficient code for processors with parallel, pipelined functional units [47, 56]. In software pipelining, an iteration of a loop in the source program is initiated before its preceding iteration is completed; thus at any time, multiple iterations are in progress simultaneously, in different stages of the computation. The steady state of this pipeline constitutes the loop body of the object code.

The drawback of software pipelining is that the problem of finding an optimal schedule is NP-complete. There have been two approaches to software pipelining: (1) change the architecture, and thus the characteristics of the constraints, so that the problem is no longer NP-complete, (2) use software heuristics. The first approach is used in the polycyclic architecture; a specialized crossbar is proposed to make optimizing loops without data dependencies between iterations tractable [47]. However, this hardware feature is expensive to build; and, when inter-iteration dependency is present within the loop, exhaustive search on the strongly connected components of the data flow graph is still necessary [31]. The second approach is used in the FPS-164 compiler [56]. However, software pipelining is applied to a limited class of loops, namely loops that contain only a single Fortran statement.

This research shows that software pipelining is a practical, efficient, and general technique for scheduling the parallelism in a VLIW machine.

We have extended previous results of software pipelining in two ways. First, we show that near-optimal results can be obtained for all loops using software heuristics. We have improved and extended previous scheduling heuristics and introduced a new optimization called *modulo variable expansion*. The latter has part of the functionality of the specialized hardware proposed in the polycyclic machine, thus allowing us to achieve similar performance.

Second, software pipelining has previously been applied only to loops with straight-line loop bodies. In this work, we propose a *hierarchical reduction* scheme whereby entire control constructs are reduced to an object similar to an operation in a basic block. With this scheme, software pipelining can be applied to arbitrarily complex loops. The significance is threefold: All innermost loops, including those containing conditional statements, can be software pipelined. If the number of iterations in the innermost loop is small, we can software pipeline the second level loop as well to obtain the full benefits of this technique. Lastly, hierarchical reduction diminishes the penalty of start-up cost of short vectors.

Analysis of a large set of user programs shows that the combination of software pipelining and hierarchical reduction is effective. Optimal performance is often obtained for many simple, classical systolic algorithms. Many new and more complex systolic programs have been developed using the compiler, and the quality of the generated code is comparable to that of hand crafted microcode.

1.3. This presentation

To establish the context for this book, we first describe the architecture of the Warp systolic array. Chapter 2 describes the design of the architecture and the rationale behind the design decisions. It also highlights the implications of the machine features on the complexity in programming the machine.

Chapter 3 presents the machine abstraction proposed for programmable systolic arrays. In this chapter, we first argue that the user must have control over the mapping of a computation onto the array if efficient code is to be obtained for the complete computation domain of a programmable, high-performance systolic array. Next, we explain the interaction between the internal timing of a cell and the efficiency of a systolic algorithm. We then compare several existing communication models and show that the asynchronous communication model is superior for both reasons of programmability and amenability to optimization. We also discuss the hardware and software support for this communication model.

The fourth chapter introduces the W2 programming language, and presents an overview of the compiler. It describes the different modules of the compiler and their interactions. The main purpose of this chapter is to prepare the ground for discussion of the cell level optimizations in the next two chapters: Chapter 5 describes software pipelining and Chapter 6 describes hierarchical reduction.

In Chapter 5, we concentrate on the technique of software pipelining, and describe the algorithm for scheduling innermost loops whose bodies consist only of a single basic block. We present the mathematical formulation of the problem, describe our scheduling heuristics, introduce a new optimization called modulo variable expansion, and compare our approach to existing hardware and software approaches to the problem.

Chapter 6 presents the hierarchical reduction technique; this technique allows us to model entire control constructs as simple operations in a basic block. This chapter shows how this approach makes list scheduling applicable across basic blocks, and software pipelining applicable to all loops. In this chapter, we also compare our overall scheduling strategy (using both software pipelining and hierarchical reduction) with other methods proposed for handling high degrees of parallelism and pipelining in the data path: trace scheduling [19] and vector processing.

Chapter 7 presents the performance data on the Warp machine and the compiler. We first present experimental data on a collection of 72 user programs. There are two parts to the evaluation: measurements on entire programs and measurements on the innermost loops of the programs. In both parts of the experiment, we compare the performance of the code against unoptimized code to study the significance of the scheduling techniques, and we also compare the performance of the code with a theoretical upper bound to measure the efficiency of the code. Second, we present the Livermore Loop benchmark results on single Warp cells. Besides analyzing the efficiency of the compiler, we also present some insights on the effectiveness of the architecture.

Lastly, Chapter 8 summarizes the results of the research and presents the conclusions of this work. We present the conclusion that the concept of systolic processing can be fruitfully applied to an array of high-performance processors.

2
Architecture of Warp

The Warp project is a study in *high-performance, programmable systolic* architectures. Although the feasibility of the concept of systolic computing has long been established, the results have mainly been theoretical in nature, and many lower-level architectural and implementation issues have not been studied and addressed. We believed that these practical issues would not even be exposed, let alone resolved, without implementation and experimentation. We did not just set out to implement a systolic array to perform a specific function, we extended the concept of systolic processing by implementing each processing unit as a programmable, high-performance processor. The tenet of the project is that the theories and results in systolic computing research are applicable also to arrays of processors that are programmable and powerful.

The Warp array represents a serious departure from the traditional systolic machine model; it is a result of a careful analysis and re-evaluation of every feature that has always been associated with systolic processing. Each processor is implemented as an individually programmable micro-engine, with a large local data memory. It achieves a high peak computational bandwidth by allowing direct access of its multiple functional units through a wide instruction word. It has a high communication bandwidth typical of systolic arrays; but more importantly it contains architectural features that allow the bandwidth to be used effectively. The Warp array does not only serve as a vehicle for implementing existing systolic algorithms, its flexibility and programmability also open up possibilities for new parallel algorithms.

The architecture of Warp has gone through several revisions: two prototype versions, both implemented on wire-wrapped boards, and the production machine implemented on printed circuit boards. In the following, we refer to the production machine as the PC machine, for short. These two architectures are followed on by iWarp, a single-chip processor with a much improved architecture. The goal of the prototype machines was to demonstrate the feasibility of such a system, and its target application domain was quite limited. The experience gained in using the prototypes enabled us to implement a production machine that can support a larger computation domain. The architecture of the integrated version is a complete redesign to take advantage of the new implementation technology. The availability of the iWarp processor would greatly reduce the size and cost, and enhance the reliability, of complete systems.

The compiler research has been instrumental in the development of the architecture of Warp. Preliminary studies of the compiler design have contributed to the design of the prototypes. Insights gained from the construction of the compiler as well as the evaluation of the compiler generated code were incorporated into the PC machine. This chapter presents the architecture of the 10-cell prototype system in detail, as this is the architecture on which the research is based. Major revisions to the prototype in designing the PC machine are also described. This chapter also discusses the application domain of the Warp architecture, as well as its programming complexity.

2.1. The architecture

Warp is integrated into a general purpose host as an attached processor. There are three major components in the system – the Warp processor array (*Warp array*), the interface unit (*IU*), and the *host*, as depicted in Figure 2-1. The Warp array performs computation-intensive routines such as low-level vision routines or matrix operations. The IU transfers data between the array and the host; it also generates addresses for local cell memories and systolic control signals used in the computation of the cells. The host supplies data to and receives results from the IU, in addition to executing the parts of the application programs that are not mapped onto the Warp array. For example, the host performs decision-making processes in robot navigation, beam adaptation in sonar processing, and evaluation of convergence criteria in iterative methods for solving systems of linear equations.

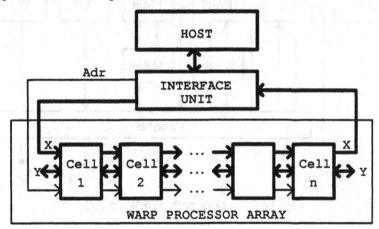

Figure 2-1: Warp system overview

The Warp array is a one-dimensional systolic array with identical cells called Warp cells. Data flow through the array on two data paths (X and Y), while addresses (for local cell memories) and systolic control signals travel on the Adr path (as shown in Figure 2-1). Each cell can transfer up to 20 million 32-bit words (80 Mbytes) per second to and from its neighboring cells, in addition to propagating 10 million 16-bit addresses across the array per second. The Y data path is bidirectional, with the direction configurable statically.

2.1.1. Warp cell

Each Warp cell is a high-performance processor, with its own sequencer and program memory. It has a wide machine instruction format: units of the data path are directly controlled through dedicated fields in the instruction. The data path of a processor in the prototype array is illustrated in Figure 2-2. It consists of a 32-bit floating-point multiplier

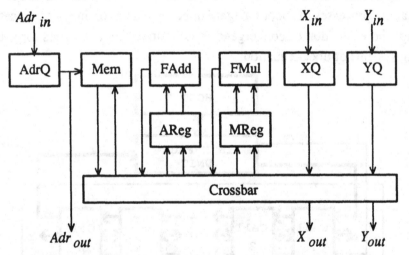

Figure 2-2: Data path of a Warp cell

(Mpy) and a 32-bit floating-point adder (Add), a local memory for resident and temporary data (Mem), a queue for each inter-cell communication path (X, Y, and Adr Queue), and a register file to buffer data for each floating-point unit (AReg and MReg). All these components are interconnected through a crossbar switch. The instructions are executed at the rate of one instruction every major clock cycle of 200 ns. The details of each component of the cell, and the differences between the prototype and the production machine, are given below.

Floating-point units. The floating-point multiplier and adder are implemented with commercial floating-point chips [59]. At the time of design for the Warp machine, these floating-point parts depended on extensive pipelining to achieve high performance. Both the adder and multiplier have 5-stage pipelines.

Data storages. The cell memory hierarchy includes a data memory and two 4-ported register files, one for each functional unit. The size of the data memory of the prototype is 4 Kwords, and that of the PC machine is 32 Kwords. In an instruction, two memory accesses can be made, one read and one write. The register files serve as buffers for the operands of the multiplier and the adder. The register files hold 31 words of data each. (Actually the register files hold 32 words; it is written to in every clock, so one word is used as a sink for those instructions without useful write operations). An instruction can read two words from and write two words into each register file. There is a two-clock delay associated with the register file. Therefore, the result of a floating-point operation cannot be used until a total of seven clock ticks later.

Address generation. The Warp cells in the prototype machine are not equipped with integer arithmetic capability. The justification is that systolic cells typically perform identical, data-independent functions, using identical addressing patterns and loop controls. For example, when multiplying two matrices, each cell is responsible for computing several columns of the results. All cells access the same local memory location, which has been loaded with different columns of one of the argument matrices. Therefore, common addresses and loop control signals can be generated externally in the IU and propagated to all the cells. Moreover, it is desirable that each Warp cell performs two memory references per instruction. To sustain this high local memory bandwidth, the cell demands a powerful address generation capability, which was expensive to provide. We could dedicate much more hardware resources to address generation by implementing it only once in the IU, and not replicated on all Warp cells. Although all cells must execute the same program and use the same addresses, they do not necessarily operate in lock step. As we shall show, cells in a systolic array often operate with a time delay between consecutive cells, The queue along the address path allows them to operate with different delays at different times.

The lack of address generation capability on the cells means that cells cannot execute different programs efficiently. The finite queue size in the address path also places a limitation on programs in which cells execute the same code, since the delay between cells is limited by the length of the queue. At the time the PC was designed, a new address generation part became available. This part can supply two addresses every clock and includes a 64-register file. As the cost of address generation capability is significantly lowered, it is provided on each cell in the PC Warp.

Inter-cell communication. Each cell communicates with its left and right neighbors through point-to-point links, two for data and one for addresses and control signals. A queue is associated with each link and is placed on the input side of each cell. This queue is 128 and 512 words deep in the prototype and the PC Warp machines, respectively. The queues are important because they provide a buffer for data communicated between cells, and thus allow the communicating cells to have different instantaneous input/output rate.

In the prototype machines, flow control is not provided in hardware. That is, the software must ensure that a cell never reads from an empty queue or writes into a full queue. Dynamic flow control is provided in the PC Warp: a cell is blocked whenever it tries to read from an empty queue or write to a full queue.

Crossbar switch. Experience with the Programmable Systolic Chip showed that the internal data bandwidth is often the bottleneck of a systolic cell [22]. In the Warp cell, the two floating-point arithmetic units can consume up to four data items and generate two results per instruction. The data storage blocks are interconnected with a crossbar to support this high data processing rate. There are six input and eight output ports connected to the crossbar switch; six data items can be transferred in a single clock, and an output port can receive any of the data items.

Control path. Each Warp cell has its own local program memory and sequencer. Even if the cells execute the same program, it is not easy to broadcast the microinstruction words to all the cells, or to propagate them from cell to cell, since the instructions are very wide. Moreover, although the cell programs may be the same, cells often do not execute them in lock step. The local sequencer also supports conditional branching efficiently. In SIMD machines, branching is achieved by masking. The execution time is equivalent to the *sum* of the execution time of both branches. With local program control, different cells may follow different branches of a conditional statement depending on their individual data; the execution time is the *maximum* of the execution time of both branches.

The data path is controlled by wide instruction words, with each component controlled by a dedicated field. In a single instruction, a processor in the Warp prototype array can initiate two floating-point operations, read and write one word of data from and to memory, read four words and write four words to the register files, receive and send two data items from and to its two neighbors, and conditionally branch to a program location. In addition, the PC machine can perform two integer operations, which include reading the operands and storing the results into the register file of the address generation unit. The orthogonal instruction organization makes scheduling easier since there is no interference in the schedule of different components.

2.1.2. Interface unit

The IU generates data memory addresses and loop control signals for the cells. In each instruction, the IU can compute up to two addresses, modify two registers, test for the end of a loop, and update a loop counter. To support complex addressing schemes for important algorithms such as FFT, the IU also contains a table of pre-stored addresses; this table can be initialized when the microcode is loaded. The IU has the ability to receive four 8-bit or two 16-bit data packed into one 32-bit word. Each data is unpacked and converted to a 32-bit floating-point data before it is sent to the Warp array. The opposite transformation is available as well.

2.1.3. The host system

The Warp host, depicted in Figure 2-3, consists of a SUN-3 workstation (the master processor) running UNIX and an external host. The workstation provides a UNIX environment to the user. The external host consists of two *cluster processors* and a *support processor*. The processors in the external host run in stand-alone mode. The support processor controls peripheral I/O devices (such as graphics boards), and handles floating-point exceptions and other interrupt signals from the Warp array. The two clusters buffer the data and results to and from the array. They work in tandem, each handling a unidirectional flow of data to or from the Warp processor, through the IU. The two clusters can exchange their roles in sending or receiving data in different phases of a computation, in a ping-pong fashion.

The external host is built around a VME bus. The two clusters and the support processor each consists of a stand-alone MC68020 microprocessor (P) and a dual-ported memory (M), which can be accessed either via a local bus or via the global VME bus. The local bus is a VSB bus in the PC Warp machine and a VMX32 bus for the prototype;

the major improvements of VSB over VMX32 are better support for arbitration and the addition of DMA-type accesses. Each cluster has a *switch* board (S) for sending and receiving data to and from the Warp array, through the IU. The switch also has a VME interface, used by the master processor to start, stop, and control the Warp array. The VME bus of the master processor inside the workstation is connected to the VME bus of the external host via a bus-coupler. The total memory in the external host can be up to 36 Mbytes.

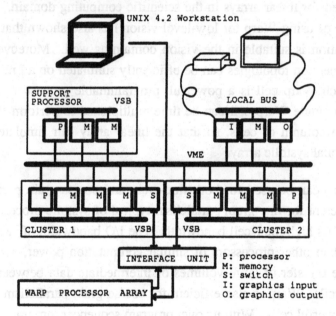

Figure 2-3: Host of the Warp machine

2.2. Application domain of Warp

The architecture of the Warp array can support various kinds of algorithms: fine-grain or coarse-grain parallelism, local or global operations, homogeneous or heterogeneous computing. (An algorithm is homogeneous if all cells execute the same program, and heterogeneous otherwise.) The factors contributing to the versatility of Warp include: simple topology of a linear array, powerful cells with local program control, large data memory and high inter-cell communication bandwidth. These features support several important problem decomposition methods [30, 37, 38].

The configuration of a linear array is easy to implement in hardware, and demands the smallest external I/O bandwidth, since only the two end-cells communicate with the outside world. Also, it is easier to use than other higher-dimensional arrays. Many algorithms have been developed for linear arrays in the scientific computing domain. Our experience of using Warp for low-level vision has also shown that a linear organization is suitable in the vision domain as well. Moreover, other interconnection topologies can be efficiently simulated on a linear array, since each Warp cell is a powerful, programmable processor. For example, a single Warp cell can be time multiplexed to perform the function of a column of cells, so that the linear array can simulate a two-dimensional systolic array.

Warp can be used for both fine-grain and coarse-grain parallelism. It is efficient for fine-grain parallelism typical of systolic processing, because of its high inter-cell bandwidth. The I/O bandwidth of each cell is higher than other processors of similar computation power, and it supports the transfer of large volumes of intermediate data between neighboring cells. Warp is also efficient for coarse-grain parallelism because of its powerful cells. With its own program sequencer, program memory and data memory, each cell is capable of operating independently. The data memory sizes (4 Kwords in the prototype machine, and 32 Kwords in the PC machine) are large for systolic array designs. It has been shown that the I/O requirement to sustain a high computation rate is lower for machines with larger data memories [35].

Systolic arrays are known to be effective for *local* operations, in which each output depends only on a small corresponding area of the input. Warp's large memory and high I/O bandwidth enable it to perform *global* operations in which each output depends on any or a large portion of the input [37]. Examples of global operations are FFT, component labeling, Hough transform, image warping, and matrix computations such as matrix multiplication or singular value decomposition (SVD). The ability to perform global operations as well significantly extends the computation domain of the machine.

2.3. Programming complexity

While the potential performance of a high-performance systolic array like Warp is enormous, the complexity of using the machine is proportionally overwhelming. Parallelism exists at several levels in the Warp architecture. At the system level, there are separate processors for input/output (the external host), control (IU), and data computation (Warp array). At the array level, there are ten cells tightly coupled together to solve a single problem; and finally at the cell level, there is a horizontal architecture with multiple pipelined functional units.

System level. Communication between the host and the processor array requires code to be generated for all three components of the system. Data must first be transferred from the host to the memory of the input cluster processor. The input processor then sends the data to the IU in the order the data is used in the Warp cells. The IU and the cluster memories communicate asynchronously via a standard bus. The IU feeds the data received from the cluster to the first Warp cell. The output process, however, requires cooperation in software only between the cell outputting the data and the host. The outputting cell tags the data with control signals for the IU output buffer, and the host reads the data directly from the buffer via the standard bus. In the prototype machines, where there is no flow control provided in the queues on the cells, the IU must send the data to the first cell before it tries to remove data from the queue. If the cluster processor cannot supply data in time, the entire Warp array is stalled.

For the prototype machine, since the cells do not have local address generation capability, they must cooperate closely with the IU. Data independent addresses and loop control signals used in the cells are computed on the IU and propagated down to the individual cells. Since addresses and loop controls are integral to any computation, the actions taken by the IU and the Warp cells are strongly coupled.

Massive amounts of detail must be mastered to use the Warp system. Fortunately, the problem of using this level of parallelism is not intrinsically hard, and can be solved through careful bookkeeping. This problem will be addressed in Chapter 4 where the overall structure of the compiler is described.

Array level. The mapping of a computation onto a locally connected array of processors has been an active area of research for the last several years. Systolic array designs require the workload be partitioned evenly across the array. Any shared data between cells must be explicitly routed through the limited interconnection in the array. The computation and the communication must be scheduled such that the cells are utilized as much as possible. Automatic synthesis techniques have been proposed only for simple application domains and simple machine models. The problem of using the array level concurrency of a high-performance array effectively is an open issue.

The approach adopted in this work is to expose this level of concurrency to the users. The user specifies the high-level problem decomposition method and the compiler handles the low-level synchronization of the cells. The justification of the approach and the exact computation model are presented in the next chapter.

Cell level. The 10 MFLOPS peak computation rate of a Warp processor is achieved through massive pipelining and parallelism. Sophisticated scheduling techniques must be employed to use this level of concurrency effectively. For example, a multiply-and-accumulate step takes 14 instructions on the Warp cell; the full potential of the cell can be

achieved only if 14 such steps are overlapped. This scheduling problem
has been studied in the context of horizontal microcode compaction [23],
array processors [44, 47, 56], and VLIW (Very Long Instruction Word)
machines [20]. Details on this level of parallelism and the proposed
scheduling techniques are described in Chapters 5 and 6.

3

A Machine Abstraction

This chapter studies the design of a machine abstraction for systolic arrays with high-performance cells. While the study is based on the Warp architecture, the design is applicable to any programmable array for which a high-level language is used to specify fine-grained parallel computation on the cells. In this chapter, we also discuss the hardware and software necessary to support the proposed machine abstraction.

The machine abstraction defines the lowest level details that are exposed to any user of the machine. Therefore, it must be both general and efficient. This chapter first argues that the user must have control over the mapping of a computation across the array to achieve both generality and efficiency. The user should specify the actions for each cell; the language in which the programs are specified, however, should be supportive with high-level constructs.

Next, we compare various communication models for expressing the interaction between cells. High-performance cells typically employ optimization techniques such as internal pipelining and parallelism. We show that if we want to achieve efficiency while hiding the complexity of the cell architecture to the user, we must also abstract away the low-level cell synchronization. We, therefore, propose to use an asynchronous communication model, not only because it is easier to use but also because it is more amenable to compiler optimizations.

Lastly, this chapter discusses the hardware and software support for the asynchronous communication model. The asynchronous communication model does not necessarily have to be supported directly in hardware. More precisely, it is not necessary to implement dynamic flow control, the capability to stall a cell when it tries to read from an empty queue, or send to a full queue. Many systolic algorithms, including all previously published ones, can be implemented on machines without dynamic flow control hardware. This chapter describes how compile-time flow control can be provided.

3.1. Previous systolic array synthesis techniques

Systolic algorithm design is nontrivial. The high computation throughput of systolic arrays is derived from the fine-grain cooperation between cells, where computation and the flow of data through the array are tightly coupled. Processors in a systolic array do not share any common memory; any common data must be routed from cell to cell through the limited interconnection in the array. This complexity in designing a systolic algorithm has motivated a lot of research in systolic array synthesis [7, 13, 26, 36, 40, 41, 46].

Automatic synthesis has been demonstrated to be possible for simple problem domains and simple machine models [6, 13, 26, 46]. The target machine model of previous systolic work was custom hardware implementation using VLSI technology. The main concerns were in mapping specific algorithms onto a regular layout of simple, identical hardware components. The computation performed by each cell must therefore be regular, repetitive and data-independent. The issues that must be addressed for generating code for a linear high-performance systolic array are significantly different. Computation amenable to automatic synthesis techniques can be mapped onto a linear array fairly easily. The issues of using an array of powerful cells are in mapping complex, irregular computations onto the linear configuration, and in allowing for the complex cell timing of a high-performance processor.

The machine model assumed by these array synthesis tools is that each cell performs one simple operation repeatedly. The operation itself is treated as a black box; the only visible characteristics are that in each clock cycle, a cell would input data, compute the result, then output the result at the end of the clock. The result is available as input to other cells in the next clock.

Computations suitable for direct mapping onto silicon must be repetitive and regular. They consist of large numbers of identical units of operations so that the array can be implemented through replication of a basic cell. The data dependencies between these units must also be regular so that the interconnection of these cells do not consume much silicon area and can be replicated easily. Examples of such computations are uniform recurrence equations [46].

The emphasis of the research was in the partitioning of computation onto a two-dimensional array. Although the notation and the techniques used in the different approaches may be different, most of them have a similar basic model: The computation is modeled as a lattice with nodes representing operations and edges representing data dependencies. Each operation node executes in unit time. The lattice is mapped onto a space-time domain; the time and space coordinates of each node indicate when and where in the array to perform the operation. This modeling of the synthesis problem is powerful for mapping a regular computation onto a regular layout of simple cells. On the other hand, these built-in assumptions of regular computation and simple cells make these synthesis techniques unsuitable for high-performance arrays such as the Warp machine.

While computation partitioning often requires insights into the application area, scheduling of cell computation and communication involves mostly tedious bookkeeping. There have been a couple of systolic design tools whose goals are to alleviate the user of this second step. The user is responsible for partitioning problems across the array and expressing algorithms in terms of a higher-level machine abstraction. These tools then convert such algorithms to programs for the low-level hardware. The user can write inefficient algorithms using a simple machine model, and the tools transform them into efficient ones for the complicated hardware.

For example, Leiserson and Saxe's *retiming lemma* gives the user the illusion that broadcasting is possible [41]. It converts all broadcasting signals to local propagation of signals. The *cut theorem* introduced by Kung and myself transforms systolic designs containing cells executing operations in single clock cycles to arrays with pipelined processors [36]. This technique will be described in more details in Section 3.2.2. A limitation common to both tools is that they are designed for algorithms whose communication pattern is constant across time.

A programmable array of powerful cells is capable of a far more general class of problems than those previously studied for systolic arrays. Cells do not have to compute the same function; the function can consist of different operations requiring different amounts of time; the communication pattern does not have to be constant across time. Mapping the more complex problems onto Warp requires techniques that exploit the semantics of the operations in a computation, which were treated as black boxes in previous approaches. Examples include the polynomial GCD computation [5] and connected component labeling [38].

As efficiency and generality are the goals of the machine abstraction, the model must allow the user to control the mapping of computation onto the processors. Higher level transformation tools that can map specific computational models efficiently can be implemented above the machine abstraction.

3.2. Comparisons of machine abstractions

While the decomposition of problem across cells is best handled by the user, the pipelining and parallelism in the data path of the cells can best be utilized by automatic scheduling techniques. (This claim is substantiated in Chapters 5 and 6.) Thus, we propose that a high-level language, complete with typical control constructs, be used to describe the cell computation. We are thus faced with the following question: given that the computation of a cell is to be described in a high-level language oblivious of the low-level timing of the cell, how should the interaction between cells be specified?

There have been several different models proposed for describing point-to-point communication between processes. (We assume that there is precisely one process per processor.) Some models require that the user specify precisely the operations that are to be performed concurrently; examples include the SIMD model (single instruction, multiple data) and the primitive systolic model, explained below.

SIMD is a well established model of computation for processor arrays with centralized sequencing control. In the SIMD model of computation, all cells in an array execute the same instruction in lock step. Communication is accomplished by shift operations; every cell outputs a data item to a neighbor and inputs an item from the neighbor on the opposite side.

The primitive systolic model is commonly used in systolic design descriptions and synthesis techniques. This model has a simple synchronous protocol. With each beat of the clock, every cell receives data from its neighbors, computes with the data, and outputs the results, which are available as input to its neighbors in the next clock. The same computation is repeatedly executed on every cell in the array. Computation in this model can be easily expressed as an SIMD program where the input, compute and output phases are iterated in a loop. However, this rigid control flow structure can be exploited in optimizing the program to allow for complex internal cell timing, as described below.

Both the SIMD and the primitive systolic models have the potential to be adopted as the communication model for Warp because the Warp architecture shares many similarities with machines using such models. Many programs implemented on the Warp prototype can be written as an SIMD program. We show below that these synchronous models are inappropriate because they are awkward for describing some computation modes that are well supported by the machine and, perhaps surprisingly, they are not amenable to compilation techniques used for high-performance cells.

3.2.1. Programmability

The difficulty in expressing a computation should only reflect the inherent mismatch between the architecture and the computation, and should not be an artifact of the machine abstraction. We have identified two useful problem partitioning methods, the *parallel* and *pipeline* schemes, that are well supported by the Warp architecture [4, 38]. The machine abstraction must support easy expression of algorithms employing these methods. Let us first describe these partitioning methods before we discuss the expressibility of the communication models for programs using these schemes.

3.2.1.1. Partitioning methods

In the parallel mode, each processor computes a subset of the results. In many cases, each processor needs only a subset of the data to produce its subset of results; sometimes it needs to see the entire data set. The former is called *input partitioning*, and the latter is called *output partitioning* [3].

In the input partitioning model, each Warp cell stores a portion of the input data and computes with it to produce a corresponding portion of the output data. Input partitioning is a simple and powerful method for accessing parallelism—most parallel machines support it in one form or another. Many of the algorithms on Warp, including most of the low-level vision routines, make use of it. In the output partitioning model, each Warp cell processes the entire input data set, but produces only part of the output. This model is used when the input to output mapping is not regular or when any input can influence any output. This model usually requires a lot of memory because either the required input data set must be stored and then processed later, or the output must be stored in memory while the input is processed, and then output later. An example of an algorithm that uses output partitioning is Hough transform [3].

In the pipelined mode, the computation is partitioned among cells, and each cell does one stage of the processing. This is the computation model normally thought of as "systolic processing." It is Warp's high inter-cell bandwidth and effectiveness in handling fine-grain parallelism that makes it possible to use this model. An example of the use of pipelining is the solution of elliptic partial differential equations using successive over-relaxation [61]. In the Warp implementation, each cell is responsible for one relaxation [3]. In raster order, each cell receives inputs from the preceding cell, performs its relaxation step, and outputs the results to the next cell. While a cell is performing the k#th relaxation step on row i, the preceding and next cells perform the $k-1$st and $k+1$st relaxation steps on rows $i+2$ and $i-2$, respectively. Thus, in one pass through the 10-cell Warp array, the above recurrence is applied ten times. This process is repeated, under control of the external host, until convergence is achieved.

3.2.1.2. Programmability of synchronous models

Consider the following example. Suppose we want to evaluate the polynomial

$$P(x)=C_m x^m+C_{m-1} x^{m-1}+ \ldots +C_0$$

for $x_1, \ldots, x_n$. By Horner's rule, the polynomial can be reformulated from a sum of powers into an alternating sequence of multiplications and additions:

$$P(x)=((C_m x+C_{m-1})x+ \ldots +C_1)x+C_0$$

The computation can be partitioned using either the parallel or pipelined model. To evaluate the polynomials according to the parallel model, each cell in the array computes $P(x)$ for different values of x. Under the pipeline model, the computation can be partitioned among the cells by allocating different terms in Horner's rule to each cell.

The description of a systolic array for solving polynomials of degree 9 in pipelined mode, using the primitive systolic array model, is given in Figure 3-1. The steady state of the computation is straightforward: In

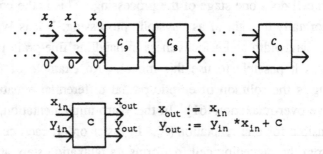

Figure 3-1: Systolic array for polynomial evaluation in pipeline mode

each clock cycle, each cell receives a pair of data, performs a multiplication and an addition, and outputs the results to the next cell. However, the boundary conditions are more complex: only the first cell is supplied with valid data in the first clock cycle, the rest of the cells must wait or compute with invalid data until the first valid one arrives. Since the first result does not emerge from the last cell until the end of the $m+1$st clock, the computation must be iterated $n+m$ times to calculate the polynomial for n sets of data, and m sets of fictitious data must be tagged on at the end of the input data.

We encounter the same problem when expressing the program in the SIMD model of computation. Using a Pascal-like notation, the program for polynomial evaluation in pipeline mode would look like the following:

```
/* shift in the coefficients */
c := 0.0;
For i := 0 to m do begin
   c := shift (R, X, c);
end;

/*  compute the polynomials  */
xdata := 0.0;
yout := 0.0;
For i := 0 to n+m do begin
   yin := shift (R, Y, yout);
   xdata := shift (R, X, xdata);
   yout := yin * xdata + c;
end;
```

The **shift** operation takes three arguments: direction, channel used, and the value to be shifted out. The direction **R**, or **L**, specifies that data is shifted from the right to the left, or left to right, respectively. The channel, **X** or **Y**, specifies the hardware communication link to be used. Lastly, the value of the third argument is the value sent to the next cell, whereas the result shifted into the cell is returned as the result of the construct. The first loop shifts in the coefficients: the first input is the coefficient for the last cell. The second loop evaluates the polynomials.

In each iteration of the second loop, results from the previous iteration are shifted out and new data for the current iteration are shifted in. Again, to compute the polynomials for n different values of x, the loop has to be iterated $n+m$ times.

The notation is equally inelegant for describing programs that use the parallel mode of decomposition if cells share common data. If the SIMD model of computation is to be used for systolic arrays, then common data must be propagated from cell to cell in the array. The program for evaluating polynomials under the parallel model is:

```
/* shift in the values of x  */
data := 0.0;
For i := 0 to n-1 do begin
   data := shift (R, X, data);
end;

/*  compute the polynomials  */
result := 0.0;
c := 0.0;
For i := 0 to n+m-1 do begin
   c := shift (R, X, c);
   result := result * data + c;
end;

/* shift out the polynomials */
For i := 0 to n-1 do begin
   result := shift (R, X, result);
end;
```

Again, the boundary conditions are complex because not all cells can start computing with meaningful data at the same time. The user must supply fictitious data to the cells and specify which of the outputs of the array constitutes the desired results.

3.2.2. Efficiency

The machine abstraction must permit automatic translation of user's input to efficient code that exploits the potential of the parallel and pipelined resources in each and every cell. It might appear that if the user synchronized the cells explicitly, less hardware or software support would be needed to generate efficient code although the programming task would be more complicated. This is unfortunately not true. Synchronization is related to the internal timing of the cells. If the user does not know the complex timing in the cells, the synchronization performed by the user is not useful. In fact, a primitive communication model not only complicates the user's task, it may also complicate the compiler's as well. Therefore, a model that abstracts away the cells' internal parallelism and pipelining should also hide the synchronization of the cells.

To illustrate this interaction between the internal timing of the cells on systolic algorithm design, let us consider implementing the algorithm in Figure 3-1 on a highly pipelined processor. Suppose the processor has a 3-stage multiplier and a 3-stage adder in its data path. The optimal throughput of an array with such processors is one result every clock cycle. This can be achieved by pipelining the computation, and inserting a 6-word buffer into the x communication path between each pair of cells. A snapshot of the computation of such an array is shown in Figure 3-2.

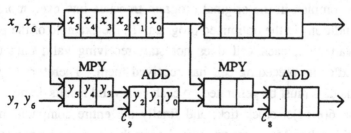

Figure 3-2: Polynomial evaluation on parallel and pipelined processors

In the original algorithm, consecutive cells process consecutive data items concurrently; in the pipelined version, by the time the second cell starts processing the ith data item, the first has already started on the $i+6$th. Therefore, the decision as to which operations should be executed in parallel must be made with the knowledge of the internal timing of the cells. This is the basic reason why synchronous models in which users specify the concurrent operatiopns do not simplify the task of an optimizing compiler.

The SIMD model of computation offers no assistance to the compiler in translating a program written with a simple machine model into one for pipelined processors. The primitive systolic array model, however, does. The regularity in the computation of systolic arrays can be exploited to allow for this second level of pipelining in the processors. Our cut theorem [36] states that pipelining can be introduced into processors of a systolic array without decreasing the throughput in terms of results per clock, if the data flow through the array is acyclic. This is achieved by adding delays on selected communication paths between the array. If results generated in one clock are used in the next by the same cell, the data flow is still considered cyclic. In case of cyclic data flow through the array, the throughput of the array can only be maintained by interleaving multiple independent computations.

Arrays such as the one in Figure 3-2 can be generated automatically by the use of the cut theorem. However, the boundary conditions that are already complex in the original program translate into even more complex conditions in the optimized program. In the example of the polynomial evaluation, each cell does not start receiving valid data until 6 clocks after its preceding cell has received them. Therefore, 6 sets of fictitious data must be generated on each cell, valid results do not emerge until the $6(m+1)$st clock tick, and finally the entire computation takes $n+6m+5$ clocks. More importantly, the cut theorem can be applied only to simple systolic algorithms, in which all cells repeat the same operation all the time. Any complication such as time-variant computation, heterogeneous cell programs, or conditional statements would render this technique inapplicable.

In summary, synchronous computation models in which the user completely specifies all timing relationships between different cells are inadequate for high-performance arrays. The reason is that they are hard to program and hard to compile into efficient code. Efficiency can be achieved only in the case of simple programs written using the primitive systolic model. If the primitive systolic model were adopted as the machine model, the versatility of the Warp machine would have been severely reduced.

3.3. Proposed abstraction: asynchronous communication

As shown by the example above, timing information on the internal cell behavior must be used to decide which operations should be executed concurrently on different cells. That is, if the machine abstraction hides the internal complexity of the cells from the user, it must also be responsible to synchronize the computations across the cells. Therefore we propose that the user programs the interaction between cells using asynchronous communication operations: Cells send and receive data to and from their neighbors through dedicated buffers. Only when a cell tries to send data to a full queue or receive from an empty queue will a cell wait for other cells.

Programs are easier to write using asynchronous communication primitives. For example, the program for evaluating polynomials is:

```
/* shift in the coefficients */
c := 0.0;
for i := 0 to m do begin
    Send (R, X, c);
    Receive (L, X, c);
end;

/*  compute the polynomials  */
for i := 0 to n-1 do begin
    Receive (L, X, xdata);
    Receive (L, Y, yin);
    Send (R, X, xdata);
    Send (R, Y, xdata * yin + c);
end;
```

Like the **shift** operation, the **receive** and **send** operations take three arguments: direction, channel used, and a variable name. In a **send** operation, the third parameter can also be an expression. The direction, **L** (left) or **R** (right), and the name of the channel, **X** or **Y**, specify the hardware communication link to be used. In a **receive** operation, the third argument is the variable to which the received value is assigned; in a **send** operation, the third argument is the value sent.

The above cell program is executed by all the cells in the array. The first loop shifts in the coefficients; the second loop computes the polynomials. In the second loop, each cell picks up a pair of **xdata** and **yin**, updates **yin**, and forwards both values to the next cell. By the definition of asynchronous communication, the computation of the second cell is blocked until the first cell sends it the first result. Figure 3-3 shows the early part of the computation of the two cells. This description is simpler and more intuitive, as the asynchronous communication model relieves the user from the task of specifying the exact operations executed concurrently on the cells. It is the compiler's responsibility to synchronize the computation of the cells correctly. This convenience in programming extends to programs using the parallel mode of problem partitioning as well.

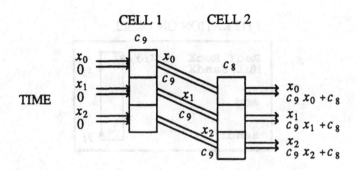

Figure 3-3: Polynomial evaluation
using the asynchronous communication model

Cell programs with unidirectional data flow written using the asynchronous communication model can be compiled into efficient array code. Cells in a unidirectional systolic array can be viewed as stages in a pipeline. The strategy used to maximize the throughput of this pipeline is to first minimize the execution time of each cell, then insert necessary buffers between the cells to smooth the flow of data through the pipeline. The use of buffering to improve the throughput has been illustrated by the polynomial evaluation example.

This approach of code optimization is supported by the high-level semantics of the asynchronous communication model. In asynchronous communication, buffering between cells is implicit. This semantics is retained throughout the cell code optimization phase, thus permitting all code motions that do not change the semantics of the computation. The necessary buffering is determined after code optimization.

3.3.1. Effect of parallelism in cells

Let us consider the compilation of the second loop in the program again for cells with a 3-stage multiplier and a 3-stage adder. In a straightforward implementation of the program, a single iteration of the polynomial evaluation loop takes 8 clocks, as illustrated in Figure 3-4.

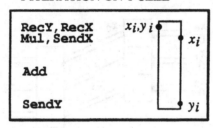

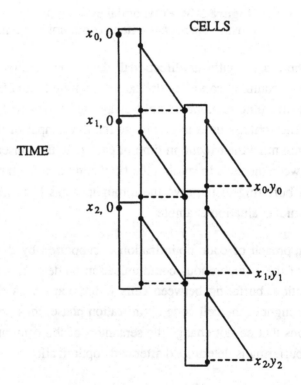

Figure 3-4: Unoptimized execution of polynomial evaluation

The figure contains the microcode for one iteration of the loop, and an illustration of the interaction between cells. The micro-operations **RecX** and **RecY** receive data from the **X** and **Y** queue in the current cell, respectively; and **SendX** and **SendY** send data to the **X** and **Y** queue of the next cell, respectively. The communication activity of each cell is captured by two time lines, one for each neighbor. The data items received or sent are marked on these lines. The solid lines connecting the time lines of neighboring cells represent data transfers on the **X** channel, whereas the dashed lines represent data transfers on the **Y** channel.

As shown in the figure, the second cell cannot start its computation until the first result is deposited into the **Y** queue. However, once a cell starts, it will not stall again, because of the equal and constant input and output rates of each cell. Therefore, the throughput of the array is one polynomial evaluation every eight clocks.

However, the hardware is capable of delivering a throughput of one result every clock. This maximum throughput can be achieved as follows: We notice that the semantics of the computation remains unchanged if we reorder communication operations on different queue buffers. This observation allows us to perform extensive code motion among the communication operations, and hence the computational operations, to achieve the compact schedule of Figure 3-5.

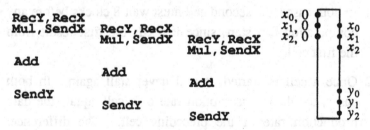

Figure 3-5: Three iterations of polynomial evaluation

The figure shows only three iterations, but this optimal rate of initiating one iteration per clock can be kept up for the entire loop using the scheduling technique described in Chapter 5. The computation of the loop is depicted in Figure 3-6. The only cost of this eight-times speed up is a longer queue between cells. While the original schedule needs a one-word queue between cells, the optimized schedule needs a six-word queue.

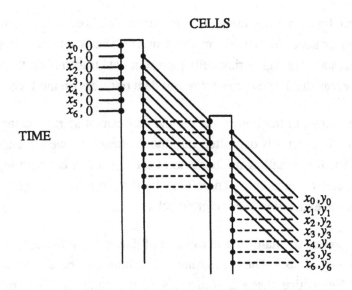

Figure 3-6: Efficient polynomial evaluation

Comparing the two programs in Figures 3-4 and 3-6, we observe the following:

1. In both cases, the second cell must wait 8 clocks before any computation can start, since it needs the first result from the first cell.

2. Once a cell is started, it will never stall again. In both cases, the data consumption rate of a cell equals the data production rate of the preceding cell. The difference, however, is that the rate in the first case is 1 in 8 instructions, whereas that in the second is 1 every instruction. This means that the latency of the first result through the array is the same, but the throughput is improved by 8 times.

3. The computations performed by the two arrays are equivalent. Since the ordering of the receive and send operations on each queue remains unchanged, the same function is performed on the data.

4. The relative ordering of operations on different queues, however, is different. Seven data items are sent on the X queue before the first output on the Y queue. Since no data is removed from the X queue until there is data on the Y queue, the X queue must be able to buffer up the seven data items. Otherwise, a deadlock situation would occur with the first cell blocked trying to send to a full X queue and the second cell blocked waiting for data on the empty Y queue. Therefore, relaxing the sequencing constraints between two queues has the effect of increasing the throughput of the system, at the expense of increasing the buffer space requirement along the communication links. The increase in buffer storage is generally insignificant with respect to the payoff in execution speed. Here, the buffer space needs only to be increased by 6 words to achieve an 8-fold speed up.

3.3.2. Scheduling constraints between receives and sends

In general, to use the internal resources of high-performance processors effectively, the sequential ordering of execution in the source program must be relaxed. The approach used in the W2 compiler is to translate the data dependencies within the computation into scheduling constraints, and allow the scheduler to rearrange the code freely so long as the scheduling constraints are satisfied. This approach supports the extensive code motion necessary to use the parallel hardware resources effectively.

As shown in the example above, communication operations in systolic programs must also be reordered to achieve efficient fine-grain parallelism. Fortunately, efficient code can be generated for unidirectional systolic programs by simply analyzing each cell independently and constraining only the ordering of communication operations *within* each cell. These sequencing constraints are represented similarly as data dependency between computational operations. The uniform representation allows general scheduling techniques to be applied to both communication and computational operations.

3.3.2.1. The problem

Communication operations cannot be arbitrarily reordered, because reordering can alter the semantics of the program as well as introduce deadlock into a program. To illustrate the former, consider the following:

(a) First cell   Second cell
    Send(R,X,1);   **Receive(L,X,c);**
    Send(R,X,2);   **Receive(L,X,d);**

(b) First cell   Second cell
    Send(R,X,1);   **Receive(L,X,d);**
    Send(R,X,2);   **Receive(L,X,c);**

Programs (a) and (b) are not equivalent, because the values of the variables c and d in the second cell are interchanged.

To illustrate deadlock, consider the following examples:

(a) First cell   Second cell
    Send(R,X,a);   **Receive(L,X,c);**
    Receive(R,X,b);   **Send(L,X,d);**

(b) First cell   Second cell
    Receive(R,X,b);   **Receive(L,X,c);**
    Send(R,X,a);   **Send(L,X,d);**

(c) First cell   Second cell
    Send(R,X,a);   **Send(L,X,d);**
    Receive(R,X,b);   **Receive(L,X,c);**

The original program (a) is deadlock-free. Reordering the communication operations as in program (b) is illegal because the two cells will be blocked forever waiting for each other's input. While program (b) deadlocks irrespective of the size in the communication buffer, program (c) deadlocks only if there is not enough buffering on the channels. In this particular example, the cells must have at least a word of buffer on each channel.

We say that the semantics of a program is preserved only if an originally deadlock-free program remains deadlock-free, and that the program computes the same results. Here we answer the following question: given that no data dependency analysis is performed across cells, what are the necessary and sufficient scheduling constraints that must be enforced within each cell to preserve the semantics of the program?

Theoretically, it is possible to allow more code motion by analyzing the data dependency across cells and imposing scheduling constraints that relate computations from different cells. However, such scheduling constraints would greatly complicate the scheduling procedure. Furthermore, since receive and send operations correspond by the order in which they are executed, any conditionally executed receive or send operations would make it impossible to analyze subsequent cell interaction. Therefore, we limit the scope of the analysis to only within a cell program.

3.3.2.2. The analysis

We separate the preservation of semantics of an asynchronous systolic program into two issues: the correctness of the computation (if the program completes), and the avoidance of introducing deadlock into a program. Here we only concentrate on the scheduling constraints pertaining to the interaction between cells; data dependency constraints within the computation in each cell are assumed to be enforced.

First, to ensure correctness, the ordering of operations on the same communication channel must be preserved. That is, we cannot change the order in which the data are sent to a queue, or received from a queue. Since receive and send operations correspond by the order in which they are executed, a pair of receives on the same queue can be permuted only if the corresponding sends on the sender cell are permuted similarly, and vice versa. Therefore, if code motions on the different cells are not coordinated, the data on each queue must be sent or received in the same order as the original. Conversely, if the ordering of operations on each communication channel is preserved, provided that the program completes, the computation is correct.

On the second issue, let us first analyze the occurrence of a deadlock under a general systolic array model. (Important special cases, such as linear or unidirectional arrays, are given below.) A systolic array is assumed to consist of locally connected cells, where each cell can only communicate with its neighbors. In a deadlock, two or more connected cells must be involved in a circular wait. Each cell involved is waiting for some action by one of the other cells. If the deadlock does not occur in the original program, then the code scheduler must have moved an operation that could have prevented the deadlock past the operation on which the cell is blocked. In other words, the operation on which the cell is blocked *depends* on the execution of some preceding operation, which must be directed at one of the cells involved in the circular wait. Therefore, the key is to first identify all such dependency relationships: a pair of operations such that the first is responsible of unblocking the second. We then insert necessary scheduling constraints to ensure that operations sharing a dependency relationship are not permuted.

If no further information is available on the topology of the array, the direction of data flow, or the size of the data queues, the scheduling constraints are strict. A cell can block either on a receive or send operation, and every receive or send operation can potentially unblock a subsequent receive or send operation. Therefore, the original sequential ordering of all communication operations must be observed.

If the communication buffers are infinite in length, then cells block only on receiving from an empty queue. The only operation that could have unblocked a cell is a send operation. Therefore, the only ordering that must be preserved is between send operations and subsequent receive operations.

The scheduling constraints can be relaxed in a linear systolic array, provided that there is no feedback from the last cell to the first cell. In a linear array, two and only two cells can be involved in a circular wait: since a cell can only wait for a neighbor one at a time, it is impossible to form a cycle with more than two cells in a linear array. Therefore the unblocking of a communication operation can only depend on receive or send operations from or to *the same cell.* The scheduling constraints above can thus be relaxed as follows: if the queues are infinite in length, send operations to a cell must not be moved below receive operations from the same cell. If the queues are finite, all the sends and receives to and from the same cell must be ordered as before. That is, receive and send operations with the right neighbor are not related to the receive and send operations with the left neighbor.

If the data flow through the array is acyclic, and if the queues are infinite, then no scheduling constraints need to be imposed between communication operations on different channels. This is true of any array topology. The reason is that cells cannot be mutually blocked, and thus there is no possibility of a deadlock. However, if the queues are finite, a cyclic dependency can be formed between the cell and any two of its neighbors. Therefore, as in the general systolic array model, all send and receive operations on every cell must be ordered as in the original program.

To summarize, the ordering of all sends to the same queue, or receives from the same queue, must be preserved. Then, depending on the information available on the topology, the queue model and the data flow, different constraints apply. The information is summarized in Table 3-1. The arrow denotes that if the left operation precedes the right operation in the source program, the ordering must be enforced. The line "send(c) → rec(c)" means that the send and receive operation directed to the same cell c must be ordered as in the source program.

General Topology

QUEUE	CYCLIC DATA FLOW	ACYCLIC DATA FLOW
Finite	rec/send → rec/send	rec/send → rec/send
Infinite	send → rec	none

Linear Array

QUEUE	CYCLIC DATA FLOW	ACYCLIC DATA FLOW
Finite	rec/send(c) → rec/send(c)	rec(c) → rec(c) send(c) → send(c)
Infinite	send(c) → rec(c)	none

Table 3-1: Constraints between receive and send operations

3.3.2.3. Implications

Both the finite and infinite queue models permit extensive code motion in the compilation of unidirectional, linear systolic array programs. Since the scheduling of receive and send operations is not constrained, these operations can be scheduled whenever the data are needed, or whenever the results are computed. As in the polynomial evaluation example, the second data set can be received and computed upon in parallel with the first set. It is not necessary to wait for the completion of the first set before starting the second. The send and receive operations are arranged to minimize the computation time on each cell. Provided that sufficient buffering is available between cells, the interaction between cells may only increase the latency for each data set, but not the throughput of the system.

The major difference between the finite and infinite queue models is that in the former, data buffering must be managed explicitly be the compiler. In the finite queue model, because we cannot increase the minimum queue size requirement, the receive operations from different queues must be ordered, and so must the send operations. In the polynomial evaluation example, the result of an iteration must be sent out before passing the data (**xdata**) of the next iteration to the next cell. To overlap different iterations, the values of **xdata** from previous iterations must be buffered internally within the cell. The number of data that needs to be buffered exactly equals the increase in minimum queue size in the infinite queue model. Therefore, while the infinite queue model automatically uses the existing communication channel for buffering, the finite queue model requires the buffering to be implemented explicitly. This buffering may be costly especially when the number of operations executed per data item is small.

The infinite queue model has been used successfully in the W2 compiler for the Warp machine. The infinite size model can be adopted because the queues on the Warp cells are quite large; they can hold up to 512 words of data. The increase in queue size requirement due to the allowed code motions is small compared to the size of the queue. The queue size has not posed any problems when cells operate with a fine granularity of parallelism. This infinite queue model is likely to be an important model for other systolic architectures for the reasons below.

First, as discussed above, the infinite queue model can generally allow more code motions than the finite queue model. Second, data buffering is an important part of fine-grain systolic algorithms. It has been used in systolic algorithms to alter the relative speeds of data streams so that data of an operation arrive at the same cell at the same time. Examples include 1-dimensional and 2-dimensional convolutions. We have also shown that buffering is useful in streamlining the computation on machines with parallel and pipelined units. Moreover, a large queue is useful in minimizing the coupling between computations on different cells; this is especially important if the execution time for each data set is data dependent. Therefore, data buffering is likely to be well supported on systolic processors. The infinite queue model uses the data buffering capability of systolic arrays effectively. Lastly, the increase in the queue size requirement can be controlled. In the current implementation of W2, only communication operations in the same innermost loop are reordered. Code motion is generally performed within a small window of iterations in the loop; the size of the window increases with the degree of internal pipelining and parallelism of the cell. It is possible to further control the increase in buffer size by limiting the code motion of communication operations to within a fixed number of iterations. The analysis and the manipulation of the scheduling constraints are no different from those of computational operations involving references to array variables.

3.3.3. Discussion on the asynchronous communication model

The asynchronous communication semantics allows systolic algorithms to be optimized easily; In this approach, each cell is individually compiled and optimized, and then the necessary buffers are inserted between cells. Results similar to those of the cut theorem can be obtained. One important difference is that while the cut theorem is applicable only to simple, regular computations, the proposed approach applies to general programs.

This approach allows us to obtain efficient code when the following properties are satisfied: the machine has a long queue buffer with respect to the grain size of parallelism, the performance criterion is throughput rather than latency, and the data flow through the array is unidirectional. If any of these properties is violated, more efficient code can probably be achieved if the internal timing of the cells is considered in the computation partitioning phase, or if all the cells in the array are scheduled together.

The effect of cyclic data flow on efficiency depends on the grain size of parallelism. Cyclic data flow has little negative effect on computation of coarse-grain parallelism, but it may induce a significant performance loss in computation of fine-grain parallelism. A common example of the former is domain decomposition, where data is exchanged between neighboring cells at the end of long computation phases. The communication phase is relatively short compared to the computation phase, and it is not essential that it is optimal. In fine-grain parallelism, data sent to other cells are constantly fed back into the same cell. To minimize the time a cell is blocked, we must analyze the dependency across cells to determine the processing time required between each pair of send and receive operations. Besides, Kung and I showed that the internal pipelining and parallelism within a cell cannot be used effectively for a single problem in arrays of cyclic flow [36]. Multiple problems must be interleaved to use the resources in a cell effectively. In the paper, we also showed that many of the problems, for which cyclic algorithms have been proposed, can be solved by a ring or torus architecture. Rings and tori are amenable to similar optimization techniques as arrays of unidirectional data flow. Therefore, the best solution is to rewrite the bidirectional algorithms with fine-grain parallelism as algorithms on rings and tori.

3.4. Hardware and software support

The semantics of asynchronous communication can be directly supported by dynamic flow control hardware. This hardware is provided in the production version of the Warp machine, but not on the prototypes. When a cell tries to read from an empty queue, it is blocked until a data item arrives. Similarly, when a cell tries to write to a full queue of a neighboring cell, the writing cell is blocked until data is removed from the full queue. Only the cell that tries to read from an empty queue or to deposit a data item into a full queue is blocked.

The asynchronous communication model allows us to extend the original computation domain of systolic arrays, and this full generality of this model can be supported only by a dynamic flow control mechanism.

However, many systolic algorithms can be implemented without dynamic flow control hardware. Constructs that are supported on such hardware include iterative statements with compile-time loop bounds, and conditional statements for which the execution time of each branch can be bounded and that the number of receive and send operations on each queue is identical for both branches. The asynchronous communication model does not preclude such programs from being implemented on cells without dynamic flow control support. Static flow control can be implemented after the cell programs have been individually optimized. In fact, this communication model is recommended even for synchronous computation because of its amenability to compiler optimizations. This section discusses the necessary hardware support and presents an efficient algorithm for implementing compile-time flow control.

3.4.1. Minimum hardware support: queues

The minimum hardware support for communication between cells is that data buffering must be provided along the communication path as a queue. While buffering has always been used in systolic computing extensively, a delay element has typically been used. A delay on a data path is an element such that input data into the element emerges as output after some fixed number of clock cycles. Delays are inserted in data paths of many systolic algorithms to modify the speed of a data stream with respect to another [34]. They have also been used to increase the throughput of a system consisting of processors with internal parallelism and pipelining [36]. Even in our earlier example of polynomial evaluation, the queue of the communication links can be implemented by a delay of 8 clocks. In fact, in an earlier design of the Warp cell, the communication buffers were implemented as *programmable delays*.

Although programmable delays have the same first-in first-out property of queues, they are different in that the rate of input must be tightly coupled with the output. If the input rate into the delay is not one data item per clock, then some of the data item emerging from the delay are invalid and the output must be filtered. A constant delay through the queue means that the timing of data generation must match exactly that of data consumption. Any data dependent control flow in the program cannot be allowed; even if the control flow is data independent, such scheduling constraints complicate the compilation task enormously. Therefore programmable delays are inadequate as communication buffers between programmable processors in an array.

Queues decouple the input process from the output process. Although the average communication rate between two communicating cells must balance, the instantaneous communication rate may be different. A sender can proceed as long as there is room on the queue when it wants to deposit its data; similarly, a receiver can compute on data from the buffer possibly deposited a long time ago until it runs out of data. The degree of coupling between the sender and receiver is dependent on the queue length. A large buffer allows the cells to receive and send data in bursts at different times. The communication buffers on the prototype and the production Warp machines are 128 and 512 words deep, respectively. This long queue length is essential to support the optimizations above, where throughput of the array is improved at the expense of buffer space.

3.4.2. Compile-time flow control

If dynamic flow control is not provided in hardware, as in the case of the Warp prototype machine, then every effort must be made to provide flow control statically. Static flow control means that we determine at compile time when it is safe to execute a receive or send operation, so the generated code does not have to test the queue status at run time. It is expensive to provide dynamic flow control in software on a highly parallel and pipelined machine; explicit testing of queue status on every queue access would have introduced many more possible branches in the computation, and made it harder to optimize the cell computation. The status of the queues is not even available on the Warp prototype machine. Compile-time flow control is implemented by mapping the asynchronous communication model onto a new computation model, the *skewed* computation model, described below.

3.4.2.1. The skewed computation model

The skewed computation model provides a simple solution to implementing static flow control. In this model, a cell waits to start its computation by the necessary amount of time to guarantee that it would not execute a receive operation before the corresponding send is executed. The delay of a cell relative to the time the preceding cell starts its computation is called the *skew*. Code generated using this model executes as fast, and requires just as much buffering, as a program executing full speed on a machine with dynamic flow control. This section introduces the computation model and its limitations; the next section presents the algorithm for finding this skew.

Let us use a simple example to illustrate the approach. Suppose the following microprogram is executed on two consecutive cells:

$$SendX_0$$
$$RecX_0$$
$$RecX_1$$
$$Add$$
$$Add$$
$$SendX_1$$

The send and receive operations are numbered so we can refer to them individually. Suppose data only flow in one direction: each cell receives data from its left and sends data to its right neighbor.

To ensure that no receive operation overtakes the corresponding send operation, we can take the following steps: first match all receive operations with the send operations, and find the maximum time difference between all matching pairs of receives and sends. This dif-ference is the minimum skew by which the receiving cell must be delayed with respect to the sender.

Let $\tau_r(n)$ and $\tau_s(n)$ be the time the nth receive or send operation is executed with respect to the beginning of the program, respectively. The minimum skew is given by:

$$\max(\tau_s(n)-\tau_r(n)), \quad \forall 0 \le n < \text{number of receives/sends}$$

Table 3-2 shows the timing of the receive and send operations in the program. The minimum skew, given by the maximum of the differences between the receive and send operations, is three.

n	τ_s	τ_r	$\tau_s - \tau_r$
0	0	1	−1
1	5	2	3
Maximum			3

Table 3-2: Receive/send timing functions and minimum skew

Table 3-3 shows how none of the input operations in the second cell precedes the corresponding send operation in the first cell if the second cell is delayed by a skew of three clocks.

An alternative approach to the skewed computation model is to delay the cells just before the stalling receive and send operations as would happen with dynamic flow control. As illustrated in Table 3-4, the same delay is incurred in this scheme. This approach does not in-crease the throughput of the machine. The skewed computation model is simpler as it needs only to calculate the total delay necessary and insert them before the computation starts. In the alternative approach, we need to calculate the individual delay for each operation and insert the delays into the code.

The skewed computation model can be used for both homogeneous

Time	First cell	Second cell
0	$SendX_0$	
1	$RecX_0$	
2	$RecX_1$	
3	Add	$SendX_0$
4	Add	$RecX_0$
5	$SendX_1$	$RecX_1$
6		Add
7		Add
8		$SendX_1$

Table 3-3: Two cells executing with minimum skew

Time	First cell	Second cell
0	$SendX_0$	$SendX_0$
1	$RecX_0$	$RecX_0$
2	$RecX_1$	
3	Add	
4	Add	
5	$SendX_1$	$RecX_1$
6		Add
7		Add
8		$SendX_1$

Table 3-4: Dynamic flow control

and heterogeneous programs. The computation of the skew does not assume that the cells are executing the same program. Programs using either the pipelined or parallel style of computation partitioning can be mapped efficiently to this model. Multiple communication paths can also be accommodated as long as the direction of data flow is the same. The skew is given by the maximum of all the minimum skews calculated for each queue. Bidirectional data flow cannot be handled in general using this model. Also, the domain in which compile-time flow control is applicable is inherently restrictive. The restrictions of compile-time flow control and the skewed computation model are explained below.

Data dependent control flow. Compile-time flow control relies on finding a bound on the execution time of the receive and send operations. Therefore, in general, it cannot be implemented if programs exhibit data dependent control flow. Conventional language constructs that have data dependent control flow include WHILE statements, FOR loops with dynamic loop bounds and conditional statements.

Of the data dependent control flow constructs, only a simple form of the conditional statement can be supported: conditional statements that do not contain loops in its branches. The branches of a conditional statement are padded to the same length. As shown in Chapter 6, operations within a conditional statement can be overlapped with operations outside. Therefore, padding the branches to the same length does not imply a significant loss in efficiency. The important point is that now the execution time of the statement remains constant no matter which of its branches are taken. Data can be received or sent within the branches; however, the number of data received or sent on each channel must be the same for both branches. Although we cannot determine at compile time which branch is taken and thus which of the receive/send operations are performed, we can compute a bound on the execution time of the operations. The *minimum* of the ith receive operation in either branch of the conditional statement is considered to be the execution time of the ith receive operation; similarly, the *maximum* of the ith send operation in either branch of the conditional statement is considered to be the execution time of the ith send operation. The minimum is used for receives and maximum is used for sends to ensure that the skew is large enough to handle the worst case.

Bidirectional communication. The skewed computation model simplifies the compile-time control flow problem but it cannot be used on bidirectional programs. Intuitively, the skew delays the receiver with respect to the sender to ensure that no receive operation overtakes the corresponding send operation. With bidirectional flow, it is possible that the second cell must be delayed with respect to the first cell for some pair

of receive and send operations, and the first cell must be delayed with
respect to the second for some other pair of operations. Therefore, idle
clock cycles cannot be introduced only at the beginning of the computation and must be inserted in the middle of the computation of both cells.

3.4.2.2. Algorithm to find the minimum skew

In the previous section, we explained the skewed computation
model by way of a simple straight-line program. We have also explained
how conditional statements can be handled just like straight-line code,
using the lower and upper bound on the execution times of the receive
and send operations. This section describes the algorithm for handling
loops.

The complexity of determining the minimum skew in iterative
programs depends on the similarity in the control structures in which the
matching receive and send statements are nested. Figure 3-7 is an example program that illustrates these two cases. Suppose this

```
        Add
        Loop 5 times:   RecX_0
                        RecX_1
                        Add
        Add
        Add
        Loop 2 times:   SendX_0
                        SendX_1
        Add
        Add
        Loop 2 times:   SendX_2
                        SendX_3
                        SendX_4
                        Add
                        Add
        Add
```

Figure 3-7: An example program containing loops

microprogram is executed on two cells, and they both receive data from
the left and send data to the right.

Table 3-5 gives the timing information on all the receive and send operations. The control structure of the first loop is similar to that of the second loop but not the third. Both the first and second loop contain two I/O statements in each iteration. So, the first and second receive statements are always matched with the first and second send statements, respectively. Since the input rate (2 every 3 clocks) is lower than that of the output (1 every clock), the maximum skew between these loops can be determined by considering only the first iteration. Conversely, if the input rate is higher, only the time difference between the receive/send operations of the last iteration needs to be considered. In the third loop, the number of sends per iteration differs from that of receives in the first loop. Therefore, a receive statement is matched to different send statements in different iterations and all combinations of matches need to be considered in determining the skew. The analysis gets even more complex with nested loops.

Number	τ_s	τ_r	$\tau_s-\tau_r$
0	18	1	17
1	19	2	17
2	20	4	16
3	21	5	16
4	24	7	17
5	25	8	17
6	26	10	16
7	29	11	18
8	30	13	17
9	31	14	17
Maximum			18

Table 3-5: Receive and send timing for program in Figure 3-7

In most programs, the receive and send control constructs are similar since they operate on similar data structures. Furthermore, it is not necessary to derive the exact minimum, a close upper bound will be sufficient. The following mathematical formulation of the problem allows us to cheaply calculate the minimum skew in the simple cases and its upper bound in the complex ones.

Identifying all the matching pairs of receives and sends is difficult. The key observation is that it is not necessary to match all pairs of receives and sends in the calculation of the minimum skew.

A receive/send *statement* in a loop corresponds to multiple receive/send *operations*. Each receive/send statement is characterized by its own timing function, τ_{r_i} or τ_{s_i}, and an execution set E_{r_i} or E_{s_i}. The timing function maps the ordinal number of a receive or send operation to the clock it is executed. The execution set is the set of ordinal numbers for which the function is valid. For example, the **RecX$_0$** statement is executed 5 times; it is responsible for the 0th, 2nd, 4th, 6th and 8th receive operations which take place in clock ticks 1, 4, 7, 10 and 13 respectively. Its timing function is therefore

$$\tau_{r_0}=1+3n/2,$$

and its execution set is

$$\{n \mid 0 \le n \le 8 \text{ and } n \bmod 2 = 0\}$$

For each pair of timing functions, τ_{r_i} and τ_{s_j}, we would like to find $\tau_{s_j} - \tau_{r_i}$ for all n that is in the intersection of the execution sets of both functions. The maximum of the differences is the minimum skew.

Finding the exact intersection of the execution sets of two functions may be difficult if their corresponding statements belong to dissimilar control structures. For these cases, instead of using the constraints defining the sets to solve for the intersection completely, we simply use the constraints to bound the difference between the two timing functions.

We characterize each receive or send statement by five vectors of k elements, where k is the number of enclosing loops. Each element of the vector characterizes an enclosing loop, with the first representing the outermost loop. The five vectors are:

$R=[r_1, \ldots ,r_k]$:   Number of iterations

$M=[m_1, \ldots ,m_k]$:   Number of receives or sends in one iteration of the loop.

$S=[s_1, \ldots ,s_k]$:   Ordinal number of the first receive or send in the loop with respect to the enclosing loop.

$L=[l_1, \ldots ,s_k]$:   Time of execution of one iteration of the loop

$T=[t_1, \ldots ,t_k]$:   Time to start the first iteration of the loop with respect to the enclosing loop.

For uniformity in notation, the receive or send operations themselves are considered a single-iteration loop. For example, in the program in Figure 3-7, all the vectors describing the operations contain two elements; the first gives information on the enclosing loop, and the second gives information on the statement itself. Therefore, the vector R characterizing r_0 is [5,1], because it is in a 5-iteration loop, and the operation is treated as a single iteration loop. The characteristic vectors for all the receive and send operations in the program are listed in Table 3-6.

Statement	R	M	S	L	T
r_0	[5, 1]	[2, 1]	[0, 0]	[3, 1]	[1, 0]
r_1	[5, 1]	[2, 1]	[0, 1]	[3, 1]	[1, 1]
s_0	[2, 1]	[2, 1]	[0, 0]	[2, 1]	[18, 0]
s_1	[2, 1]	[2, 1]	[0, 1]	[2, 1]	[18, 1]
s_2	[2, 1]	[3, 1]	[4, 0]	[5, 1]	[24, 0]
s_3	[2, 1]	[3, 1]	[4, 1]	[5, 1]	[24, 1]
s_4	[2, 1]	[3, 1]	[4, 2]	[5, 1]	[24, 2]

Table 3-6: Vectors characterizing receive and send operations in Figure 3-7

The timing function of each statement is:

$$\tau(n)=t_1+\left\lfloor\frac{n-s_1}{m_1}\right\rfloor l_1+t_2+\left\lfloor\frac{(n-s_1)\bmod m_1-s_2}{m_2}\right\rfloor l_2+\cdots$$

Every loop nesting of the statement contributes a term to this function.
Each term consists of two parts: the starting time of the loop with respect
to its enclosing loop, and the time for executing all the iterations that
come before the one the nth receive/send is in.

By defining

$$g(j)=\begin{cases} n, & j=0 \\ (g(j-1)-s_{j-1})\bmod m_{j-1}, & \text{otherwise} \end{cases}$$

we get

$$\tau(n)=\sum_{j=1}^{k}\left(t_j+\left\lfloor\frac{g(j)-s_j}{m_j}\right\rfloor l_j\right)$$

$$=\sum_{j=1}^{k}t_j+\sum_{j=1}^{k}\left(\frac{g(j)-s_j}{m_j}l_j-\frac{(g(j)-s_j)\bmod m_j}{m_j}l_j\right)$$

$$=\sum_{j=1}^{k}t_j-\sum_{j=1}^{k}\frac{l_j}{m_j}s_j+\sum_{j=1}^{k}\frac{l_j}{m_j}(g(j)-g(j+1))$$

$$=\sum_{j=1}^{k}t_j-\sum_{j=1}^{k}\frac{l_j}{m_j}s_j+\frac{l_1}{m_1}g(1)+\sum_{j=2}^{k}\left(\frac{l_j}{m_j}-\frac{l_{j-1}}{m_{j-1}}\right)g(j)-\frac{l_k}{m_k}g(k+1)$$

The constraints defining the execution set are:

$$\sum_{i=j}^{k}s_i\leq g(j)\leq(r_j-1)m_j+\sum_{i=j}^{k}s_i$$

The timing functions for the example program and their execution
set constraints are given in Table 3-7.

The execution sets of a pair of timing functions can be disjoint, or
one may be contained in another or they may overlap partially. The
following gives an example in each category from the program in Figure
3-7.

$\tau(n)$	Function	Execution set constraints
r_0	$1+3n/2-n/2 \bmod 2$	$0\le n\le 8$ and $n \bmod 2=0$
r_1	$1+3n/2-n/2 \bmod 2$	$1\le n\le 9$ and $n \bmod 2=1$
s_0	$18+n+0n \bmod 2$	$0\le n\le 2$ and $n \bmod 2=0$
s_1	$18+n+0n \bmod 2$	$1\le n\le 3$ and $n \bmod 2=1$
s_2	$52/3+5n/3-2(n-4)/3 \bmod 3$	$4\le n\le 7$ and $(n-4) \bmod 3=0$
s_3	$52/3+5n/3-2(n-4)/3 \bmod 3$	$5\le n\le 8$ and $(n-4) \bmod 3=1$
s_4	$52/3+5n/3-2(n-4)/3 \bmod 3$	$6\le n\le 9$ and $(n-4) \bmod 3=2$

Table 3-7: Timing functions for program in Figure 3-7

Disjoint: The execution sets of the functions $\tau_{r_0}(n)$ and $\tau_{s_1}(n)$ do not intersect, since $n \bmod 2=0$ and $n \bmod 2=1$ cannot be satisfied simultaneously. That is, no instance of data items produced by s_1 is read by r_0.

Subset relationship: The execution set of $\tau_{s_0}(n)$ is completely contained in that of $\tau_{r_0}(n)$. That is, all the data items produced by s_0 are read by r_0. The time difference is given by

$$\max \tau_{s_0}(n)-\tau_{r_0}(n) = 17-n/2+n/2 \bmod 2, \text{ where } 0\le n\le 2 \text{ and } n \bmod 2=0$$
$$\le 17$$

Partially overlapping: The execution sets of $\tau_{r_0}(n)$ and $\tau_{s_4}(n)$ intersect, but are not completely overlapped. Only some of the data produced by s_4 are read by r_0. Here, instead of solving the intersection completely, which may be expensive, we use the constraints defining the sets to bound their time differences:

$$\max \tau_{s_4}(n)-\tau_{r_0}(n) = 52/3-1+(5/3-3/2)n-2(n-4)/3 \bmod 3+n/2 \bmod 2,$$
$$\text{where } 6\le n\le 8, n \bmod 2=1 \text{ and } (n-4) \bmod 3=2$$

$$\le 49/3+1/6\times 8-2/3\times 0+1/2\times 0$$

$$=17+2/3$$

Although the maximum time difference must be determined for every pair of receive and send statements, timing functions corresponding to statements in the same loop share many common terms which need to be computed only once. Also, the branch and bound technique can be applied: bounds on the timing of all the receive/send operations in the same loop can be cheaply obtained to reduce the number of pairs of functions that needs to be evaluated.

Determining the minimum buffer size for the queues is similar to determining the minimum skew. In the minimum skew problem, we define a function for each receive/send statement that maps the ordinal number of the I/O operation to time. In the minimum buffer size problem, we define a function for each receive/send statement that maps time to the number of receive and send operations.

3.4.2.3. Hardware design

Without the need to provide dynamic flow control, the design of the hardware is much simpler. However, there is one important detail: The control for inputting a data item must be provided by the sender. That is, the sender must tag the input data word with a control signal instructing the receiving cell's queue to accept the data. In an earlier prototype of the Warp cell, input data were latched under the microinstruction control of the receiving cell. That is, as the sender presents its data on the communication channel, the receiver issues the control to latch in the input in the same clock cycle. Contrary to our original belief that the coupling between the sender and receiver was no more restrictive than compile-time flow control, it could lead to an intolerable increase in code size.

In the above discussion of compile-time flow control, it is assumed that the control for latching in input is sent with the output data. If the cell receiving the data were to provide the input signals, we need to add a **LatchX** operation in the microprogram for every **SendX** operation executed by the preceding cell, at exactly the clock the operation takes

place. The simple straight-line program in Figure 3-2 would be as follows:

LatchX$_0$,   .
                                     .
                                     .
                       SendX$_0$.
                       RecX$_0$.
LatchX$_1$,   RecX$_1$.
                       Add   .
                       Add   .
                       SendX$_1$.

Each line in the program is an micro-instruction; the first column contains the **LatchX** operations to match the **SendX** operations of the output cell, and the second column contains the original program.

Since the input sequence follows the control flow of the sender, each
cell actually has to execute two processes: the input process, and the
original computation process of its own. If the programs on communicating cells are different, the input process and the cell's own computation process will obviously be different. Even if the cell programs
are identical, we may need to delay the cell's computation process with
respect to the input process because of flow control. As a result, we may
need to merge control constructs from different parts of the program.
Merging two identical loops, with an offset between their initiation
times, requires loop unrolling and can result in a three-fold increase in
code length. Figure 3-8 illustrates this increase in code length when
merging two identical loops of n iterations.

If two iterative statements of different lengths are overlapped, then
the resulting code size can be of the order of the least common multiple
of their lengths. In Figure 3-9, a loop of $3n$ iterations and a 2-instruction
loop body is merged with a loop of $2n$ iterations and a 3-instruction loop
body. The merged program is a loop of n iterations and a 6-instruction
loop body; 6 is the minimum number of clocks before the sequence of
operations repeats itself.

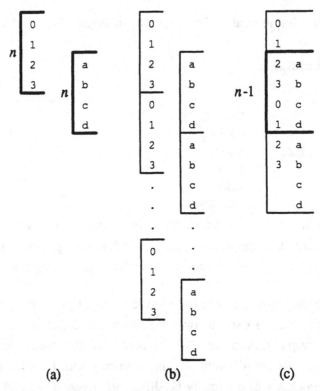

(a)   (b)   (c)

Figure 3-8: Merging two identical iterative processes with an offset
(a) original programs, (b) execution trace, and
(c) merged program

3.5. Chapter summary

The proposed machine abstraction for a high-performance systolic array is as follows: The user fully controls the partitioning of computation across the array; he sees the machine as an array of simple processors communicating through asynchronous communication primitives. The array level parallelism is exposed to the user because automatic, effective problem decomposition can be achieved only for a limited computation domain presently. On the other hand, the cell level parallelism is hidden because compiler techniques are much more effective than hand coding when it comes to generating microcode for highly parallel and pipelined processors.

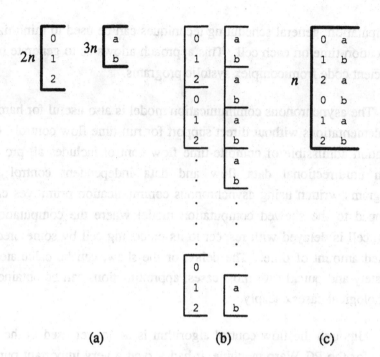

Figure 3-9: Merging two different iterative processes
(a) original programs, (b) execution trace, and
(c) merged program

The asynchronous communication model is a well-known concept in general computing, but has not been applied to systolic arrays. It is proposed here as the communication model for systolic arrays of high-performance processors not only because of its programmability but also for its efficiency. To achieve efficiency, timing information on the internal cell behavior must be used to decide which operations should be executed concurrently on different cells. That is, if the machine abstraction hides the internal complexity of the cells from the user, it must also be responsible to synchronize the computations across the cells.

The high-level semantics of the asynchronous communication model allows us to relax the sequencing constraints between the communication operations within a cell. Representing only those constraints that must be satisfied in a manner similar to data dependency constraints within a computation, general scheduling techniques can be used to minimize the execution time on each cell. This approach allows us to generate highly efficient code from complex systolic programs.

The asynchronous communication model is also useful for hardware implementations without direct support for run-time flow control. Computation admissible of compile-time flow control includes all programs with unidirectional data flow and data independent control flow. Programs written using asynchronous communication primitives can be mapped to the skewed computation model where the computation on each cell is delayed with respect to its preceding cell by some predetermined amount of time. The delay, or the skew, can be calculated accurately and quickly for most cases; approximations can be obtained for pathological cases cheaply.

Although the flow control algorithm is no longer used in the compiler for the PC Warp machine, it had served a very important purpose. It was instrumental in the development of the PC Warp architecture. We chose to use the asynchronous communication model on the prototypes because of its programmability and efficiency; it was not a decision driven by hardware. The compile-time flow control made it possible to implement this model on the prototypes, and the success in using the model led to the inclusion of run-time flow control on the PC Warp machine. A large number of applications have been developed on the prototype machine; these programs were ported to the production machine simply by recompiling the programs. The compile-time flow control algorithm may also be used in silicon compilers where custom hardware is built for specific applications.

4

The W2 Language and Compiler

The machine abstraction proposed in the last chapter is supported by the W2 language. W2 is a language in which each cell in the Warp array is individually programmed in a Pascal-like notation, and cells communicate with their left and right neighbors via asynchronous communication primitives. The user controls the array level concurrency, but not the system and cell level concurrency.

The W2 compiler is unconventional in two ways. First, code must be generated for multiple processors cooperating closely to solve a single problem. Specifically, computations on the Warp cells are tightly coupled, and data address computation and host communication must be extracted from the user's program and implemented on the IU and the host. Second, high level language constructs are directly translated into horizontal instruction words. To achieve the global code motions necessary to use the resources effectively, a good global analyzer and scheduler are essential.

This chapter presents an overview of the compiler and prepares the ground for discussion on code scheduling in the next two chapters. It first introduces the W2 language and the high level organization of the compiler, and describes how the system level of parallelism in the Warp architecture is managed by three tightly coupled code generators. The scheduling techniques used in the compiler are based on list scheduling. To introduce the notation used later, the chapter concludes with a description of the list scheduling technique as applied to the problem of scheduling a basic block.

4.1. The W2 language

The W2 language is a simple block-structured language with simple data types. We keep the language simple because we wish to concentrate our effort on efficiency. Shown in Figure 4-1 is an example of a W2 program that performs a 10×10 matrix multiplication. In the matrix multiplication program, each cell computes one column of the result. The program first loads each cell with a column of the second matrix operand, then streams the first matrix in row by row. As each row passes through the array, each cell accumulates the result for a column; at the end of each input row, the cells send the entire row of results to the host.

A W2 program is a *module*; a module has a name, a list of *module parameters* and their type specifications, and one or more *cellprograms*. The module parameters are like the formal parameters of a function; they define the formal names of the input and output to and from the array. An application program on the host invokes a compiled W2 program by a simple function call, supplying as actual parameters variables in the application program.

A cellprogram describes the action of a group of one or more cells. Although the same program is shared by a group of cells, it does not necessarily mean that all the cells execute the same instruction at the same time. As discussed in the previous chapter, computations on dif-

```
module MatrixMultiply (A in, B in, C out)
float A[10,10], B[10,10], C[10,10];

cellprogram (CellId : 0 : 9)
begin
    procedure main();
    begin
        float col[10], row, element, temp;
        int i,j;

        /* first load a column of B in each cell */
        for i := 0 to 9 do begin
            receive (L, X, col[i], B[i,0]);
            for j := 1 to 9 do begin
                receive (L, X, temp, B[i,j]);
                send (R, X, temp);
            end;
            send (R, X, 0.0);
        end;

        /* compute a row of C in each iteration */
        for i := 0 to 9 do begin
            /* each cell computes the dot product
               of its column and same row of A    */
            row := 0.0;
            for j := 0 to 9 do begin
                receive (L, X, element, A[i,j]);
                send (R, X, element);
                row := row + element * col[j];
            end;
            /*  send the result out */
            receive (L, Y, temp, 0.0);
            for j := 0 to 8 do begin
                receive (L, Y, temp, 0.0);
                send (R, Y, temp, C[i,j]);
            end;
            send (R, Y, row, C[i,9]);
        end;
    end;
end
```

Figure 4-1: W2 program for 10×10 matrix multiplication

ferent cells are typically skewed in a pipelined fashion, since a cell cannot start its computation until it has received its input data from the
preceding cell. While multiple cellprograms can be specified in the case
of the PC machine, only one cellprogram is allowed for the prototype.
This restriction is imposed by the lack of address generation capability
on the cells.   A cellprogram contains definitions of one or more
procedures.

Within a procedure, four types of statements are supported: the assignment, communication, conditional and iterative statements. The assignment, conditional and iterative statements all have conventional syntax and semantics.   However, as explained in Section 3.4.2.1, only
limited forms of the control constructs can be supported on the prototype
machines because of its lack of dynamic flow control support: Conditional statements must not contain loops in their branches, and the only
iterative statements allowed are FOR statements with compile-time loop
bounds. Such restrictions are removed for the PC machine.

There are two types of communication statements: *receive* and *send*.
They are used to specify the interaction among the cells, as well as between the host and the end cells of the array. The receive and send
statements have four parameters: the direction of the channel, the channel name, a local variable and an external (host) variable. The external
variable must be a module parameter. (A module parameter cannot be
used anywhere else.) A receive statement retrieves a data item from the
specified channel and stores it into the local variable. The first cell of the
Warp array receives data directly from the host through the IU, and the
value is explicitly specified by the external variable; all other cells
receive the data transferred in the corresponding send operation of the
communicating cell. Similarly, a send statement sends the value of the
local variable on the specified channel. In addition, the result from the
end cell is stored into the external variable on the host. This external
parameter is optional for the send statement. If no external variable is
specified, the result is not stored on the host. For example, common data

sent by the host and propagated to all the cells do not need to be stored back on the host.

4.2. Compiler overview

The compiler consists of two major phases: a machine-independent front end and a machine-dependent back end. The front end translates a W2 source program into a machine-independent flow graph and the back end translates the flow graph into code for the Warp cells, the interface unit and the cluster processors. Figure 4-2 is a diagram of the organization of the various components of the compiler.

The steps in the front end include: parsing, local data flow analysis, global data flow analysis, and machine-independent flow graph optimization. The optimizations implemented include common subexpression elimination, constant folding, height reduction, dead code removal, and idempotent operation removal [1].

Global flow analysis provides the information essential for scheduling techniques that overlap operations from different basic blocks. We call these techniques *global scheduling* techniques. They are especially important for heavily pipelined and horizontal processors because of the limited parallelism in a basic block. They rely on accurate global data dependency information. A sophisticated global flow analyzer has been implemented that generates flow information accurate up to the level of individual array elements. It analyzes the data dependency between all array accesses throughout the program, in different basic blocks, different iterations of the same loop, and across different loops. The data dependency information derived is captured by labeled arcs in the flow graph; information relevant to different code optimizations can be easily extracted and obtained from this representation.

The back end has four components: the computation decomposition unit and three code generators, for the Warp array, the interface unit and

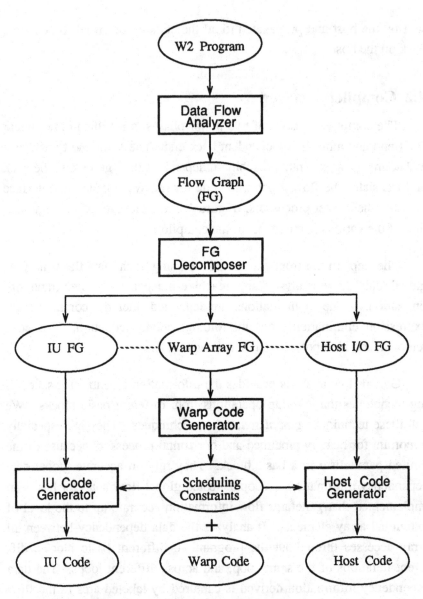

Figure 4-2: Structure of the compiler

the cluster processors. The flow graph generated by the front end is first partitioned into three subgraphs, one for each code generator. For example, a receive statement is decomposed into three parts: the host sends the data to the IU, the IU writes the data into the first cell's queue, and the cell retrieves the data item from the queue and stores it into a local variable. Also, for the prototype machines, local memory addresses determined to be data independent are separated from the flow graph. All array references indexed only by expressions of loop indices are considered data independent and are calculated on the IU. These address calculations in the cell programs are replaced by receive operations from the address queues.

The subgraph for the Warp cells is first fed to the Warp array code generator. From the code generated for the cells, the timing and sequencing information for the input to and output from the array (including addresses on the address queue) are extracted and used by the IU and the host code generators.

The code generator for the Warp cell translates the machine-independent flow graph into microcode for the cells. The process consists of the following steps:

1. Transform the machine-independent flow graph produced by the front end into a machine-dependent flow graph, where generic operators in the former are mapped onto micro-operations. This step is straightforward for Warp because of its orthogonal instruction set and the simple one-to-one correspondence between generic and machine operators.

2. Schedule the operations. The details of this step are described in the following two chapters. Chapter 5 describes the technique of software pipelining for generating highly compact code for innermost loops. Chapter 6 describes an approach called hierarchical reduction which

allows the same scheduling techniques be applied within
and across basic blocks.

3. Assign registers. As the assignment of registers is in-
   fluenced by the schedule, it is performed after code
   scheduling. The compiler uses a simple greedy algorithm
   in which registers of maximum lifetimes are assigned first.

4. For the prototype machine, calculate the minimum skew
   between cells using the algorithm explained in the preceding chapter.

The IU code generator takes its portion of the flow graph and the
timing information from the cell code generator and produces microcode
for the IU. As dynamic flow control is not implemented on the prototype
Warp machines, the IU must supply the data and addresses to the first
Warp cell before the latter attempts to remove any data item from the
queues. Therefore, the schedule of the cells imposes firm deadlines on
the IU. Various optimizations have been implemented on the IU to meet
the stringent requirements [28].

Lastly, the host code generator produces C code for the cluster
processors and the UNIX host from its portion of the flow graph. The
UNIX host transfers data between the host machine and the memory of
the cluster processors; the cluster processors transfer data between its
memory and the Warp array in the order expected by the cells. An application on the UNIX host interfaces with a program on the Warp array
simply by making C function calls.

With this overview of the compiler, we are now ready to present
some of the background material necessary for the discussion of the
scheduling techniques in the next two chapters.

4.3. Scheduling a basic block

The global scheduling techniques presented in the next two chapters use the basic framework of the list scheduling algorithm. Here, we introduce the list scheduling technique through the simpler problem of scheduling a basic block.

4.3.1. Problem definition

The objects to be scheduled in a basic block are called *micro-operation sequences*. A micro-operation sequence is a series of steps, each of which consists of zero or more micro-operations. It has the following properties:

1. Each step of the sequence may use zero or more units of resources in the machine.

2. A micro-operation sequence is atomic; once a sequence is initiated, it must run to completion without interruption. In Warp, control for the various stages of a pipelined operation is unbuffered; control must be supplied to the particular functional units at different time steps. For example, the destination of the result of an addition must be supplied exactly 7 clock ticks after the operator and the operands are specified.

3. Micro-operation sequences are minimally indivisible. They are as short as possible to maximize flexibility in scheduling. Micro-operations are grouped together as a sequence if and only if the relative timing between operations cannot be changed due to machine characteristics.

There are two kinds of scheduling constraints, resource and precedence constraints.

Resource constraints. Each step in a micro-operation sequence uses zero or more units of each kind of machine resources. Micro-operation sequences may execute concurrently only if the combined resource requirement for any one step does not exceed the available resources.

Precedence constraints. Data dependencies relating the nodes in the original machine-independent flow graph are mapped onto precedence constraints between the corresponding micro-operation sequences in the machine-dependent flow graph. Associated with each precedence constraint is a minimum delay separating the initiation of the micro-operation sequences. Two factors are involved in computing this delay: the step in which the data is accessed and the latency of the access operation. The latter depends on the machine characteristics; for example, a data item stored into memory may not be available immediately. Therefore, if the ith step of one sequence accesses the data stored in the jth step of another sequence, and that the latency of the store operation is d clock ticks, then the initiation of the first sequence must be delayed with respect to the second by $j-i+d$ clocks.

A basic block is modeled as a graph; the nodes represent the atomic micro-operation sequences, and the edges represent the precedence constraints between the sequences. The schedule of a node is the time the corresponding micro-operation sequence is initiated. The definition of the basic block scheduling problem is:

Let R be the resource configuration vector, where $R(i)$ is the number of units of resource i available in the system configuration. A resource usage vector r is a tuple of the same length where $0 \leq r(i) \leq R(i)$.

Let $G=(V,E)$ be a directed acyclic graph. Associated with each edge e is a minimum delay $d(e)$. Associated with each vertex v are the following attributes:
- length $l(v)$, and
- a set of resource usage functions $\rho_v(i)$, where

$$\rho_v(i) = \begin{cases} \text{resource usage vector of step } i, & 0 \le i < l(v) \\ <0,0,\ldots,0>, & \text{otherwise.} \end{cases}$$

The basic block scheduling problem is to find the shortest schedule $\sigma : V \to N$ such that

$$\forall e=(u,v) \in E, \ \sigma(v)-\sigma(u) \ge d(e)$$

and

$$\forall i, \ \sum_{v \in V} \rho_v(i-\sigma(v)) \le R,$$

where the length of the schedule is defined as $\max_{v \in V} \sigma(v)+l(v)$.

The problem has been shown to be NP-complete [17] by reduction from the discrete processor scheduling problem [8].

4.3.2. List scheduling

Local microcode compaction is the problem of fitting micro-operations in a basic block into as compact a schedule as possible. Various algorithms have been proposed and studied, and this problem has been considered solved [23]. A comprehensive evaluation of the different techniques has been reported in Fisher's dissertation [17]. Fisher recommends list scheduling be used because of its effectiveness and ease of implementation. Although the nodes to be scheduled execute in unit time in Fisher's experiments, list scheduling has also been reported to be effective for operations that consume multiple units of execution time [56].

The original list scheduling algorithm is slightly modified here to handle the multi-step operation sequences. During the scheduling process, the algorithm maintains a "ready" list of nodes whose predecessors have already been scheduled, and thus are ready to be scheduled themselves. The process is iterative. Each iteration consists of the following steps: select a node from the ready list according to some priority function, schedule it and then update the list. The process terminates when all the nodes have been scheduled.

The algorithm is presented in Figure 4-3. To schedule a node, first compute the earliest time the node can be initiated without violating any precedence constraints with those already scheduled. Starting from this time slot, we test successive time slots to find the first one that satisfies the resource constraints imposed by the nodes scheduled so far. The function *SatisfyResourceConstraint* can potentially be expensive; to test if a slot is eligible to start a sequence, every step in the sequence must be tested for resource conflict. To minimize the cost, the cumulative resource usage for the nodes already scheduled is incrementally compiled. By using a bit-vector array representation as suggested in Fisher's work, testing for conflicts is reduced to a series of logical operations.

4.3.3. Ordering and priority function

List scheduling offers a framework in which to attack the scheduling problem; the algorithm can be adapted and tuned to particular scheduling problems by varying the ordering in which the nodes are scheduled. In this case of scheduling a basic block, the nodes in the flow graph are traversed in a top-down manner, and the priority function used to discriminate between nodes that are ready to be scheduled at the same time is the height of the node, defined below.

By a top-down traversal of the nodes, we mean that those nodes that must execute first are scheduled last. In mapping the operations in the machine-independent flow graph to machine-dependent operations, the direction of the data dependency relationships are reversed when mapped to precedence constraints in the machine-dependent flow graph. That is, when an operation u in the machine-independent flow graph uses the result of the operation v, we say that the machine-operation representing

---

FUNCTION *ListSchedule* (V, E)
BEGIN
    Ready := root nodes of V;
    Sched := $\varnothing$;
    WHILE *Ready* $\neq \varnothing$ DO
    BEGIN
        v := highest priority node in *Ready*;
        Lb := *SatisfyPrecedenceConstraints* $(v, Sched, \sigma)$;
        $\sigma(v)$:= *SatisfyResourceConstraints* $(v, Sched, \sigma, Lb)$;
        Sched := *Sched* $\cup \{v\}$;
        Ready := *Ready*$-\{v\}+\{u\,|\,u \notin Sched \wedge \forall (w,u) \in E, w \in Sched\}$;
    END;
    RETURN (σ);
END;

FUNCTION *SatisfyPrecedenceConstraint* $(v, Sched, \sigma)$
BEGIN
    RETURN $(\max_{u \in Sched} \sigma(u)+d(u,v))$
END;

FUNCTION *SatisfyResourceConstraint* $(v, Sched, \sigma, Lb)$
BEGIN
    FOR $i := Lb$ TO ∞ DO
        IF $\forall 0 \leq j < l(v),\ \rho_v(j)+ \sum_{u \in Sched} \rho_u(i+j-\sigma(u)) \leq R$ THEN
           RETURN (i);
END;

Figure 4-3: List scheduling for a basic block

u precedes that representing u. The steps in the micro-operation sequences are inverted, and so is the final schedule generated by the algorithm; the first slot in the schedule actually corresponds to the last
microinstruction in the program. By trying to schedule the operations as
early as possible in the (inverted) schedule, the operations are actually
delayed as much as possible.

The reason for scheduling the original flow graph in an order opposite to the execution order is to minimize the lifetime of register values, the duration from the time a value is stored into a register to the last use of the value. In this reverse ordering, a store into a register is scheduled after all its uses have been scheduled. By placing the store operation as close as possible to the first use of the value, the lifetime of the value is shortened. If scheduling were performed in the order of the execution, a data value would be stored into the register as soon as the value was available regardless of when the register was referenced. The lifetimes of the values and thus the need for registers would be significantly increased.

The priority function used in the W2 compiler is rather conventional. The priority of a node is given by its height, where the height of a node is defined as the longest path from the node to a terminal node. Using this priority function, operations followed by a long chain of computation is started as soon as possible. The rationale is that computation time is governed by the critical path in the graph since the height of the graph is typically much greater than its breadth due to the deep pipelining in the arithmetic units of the machine. In previous experiments with list scheduling, priority functions that also consider the likelihood of resource contention usually performed quite well. In the case of Warp, the resource usage of each micro-operation sequence within a basic block is fairly light, and it is not included in the priority function.

In the next two chapters, the same list scheduling framework is used to schedule operations across basic blocks. However, the ordering and priority function used are quite different to adapt to the particular scheduling problems.

5

Software Pipelining

Pipelining and parallel functional units are common optimization techniques used in high-performance processors. Traditionally, this parallelism internal to the data path of a processor is only available to the microcode programmer, and the problems of minimizing the execution time of the microcode within and across basic blocks are known as local and global compaction, respectively. The development of the global compaction technique, trace scheduling, has led to the introduction of VLIW (very long instruction word) architectures [9, 19, 20, 21]. A VLIW machine is like a horizontally microcoded machine: it consists of parallel functional units, each of which can be independently controlled through dedicated fields in a "very long" instruction. A characteristic distinctive of VLIW architectures is that these long instructions are the machine instructions. There is no additional layer of interpretation where machine instructions are expanded into micro-instructions. A compiler directly generates these long machine instructions from programs written in a high-level language. A VLIW machine generally has an orthogonal instruction set; whereas in a typical horizontally microcoded engine, complex resource or field conflicts exist between functionally independent operations. Without the code selection problem of horizontally microcoded machines, the scheduling problem of the VLIW machines is significantly simplified.

A Warp cell is a VLIW machine; it has a wide instruction format and consists of multiple pipelined functional units. For example, the data path can support the overlapped execution of fourteen multiply-and-accumulate steps. In a single instruction, the Warp processor can initiate two floating-point operations, read and write a word from and into memory, input and output two data words from and to neighboring cells, branch to any location in program memory, and receive and forward two data addresses in the case of the prototype machine, and perform two integer operations in the case of the PC machine.

The VLIW machine organization is confined to the processor level architecture. The Warp machine is an MIMD (multiple instruction, multiple data) array: each cell has its own sequencing control and executes its own program. The degree of parallelism in each cell is limited to several different (pipelined) functional units. One control, or branch, operation per instruction is sufficient to balance the number of arithmetic operations that can be executed in parallel.

It is interesting to contrast our approach with the large-scale VLIW approach, where the processors contain tens of functional units and a multi-way branching capability [18]. The modular design of Warp offers a path to a scalable, high-performance system by interconnecting large number of processors together. With the support of high-bandwidth and low-latency communication, the Warp processors can be programmed to cooperate with a fine granularity of parallelism, much like the functional units in a large-scale VLIW machine. However, the individual sequencing control on the cells enable them to operate independently. This additional capability to implement coarse-grain parallel programs greatly adds to the versatility of the Warp system.

In this chapter, we propose an alternative approach to trace scheduling for scheduling VLIW machine code. We use *software pipelining* and *hierarchical reduction*. A major distinction between our approach and trace scheduling is that we exploit the characteristics of different control constructs using different techniques. *Software pipelining* is a scheduling technique that exploits the repetitive nature of loops to generate highly efficient code [44, 47, 56]. Iterative constructs deserve special attention because most computation time is spent in innermost loops, especially in numerical processing. Conditional statements, on the other hand, are difficult to optimize because less information is known about their branch probability, so we concentrate on preventing conditional statements from serializing the execution of other operations. In particular, the presence of conditional statements in a loop must not prevent the loop from being software pipelined. We propose a *hierarchical reduction* scheme whereby entire control constructs are reduced to an object similar to an operation in a basic block. This allows techniques defined for scheduling basic blocks to be applicable across basic blocks. This chapter describes software pipelining and the next chapter describes hierarchical reduction.

In software pipelining, iterations of a loop in the source program are continuously initiated at constant intervals, before the preceding iterations complete. The advantage of software pipelining is that optimal performance can be achieved with a compact object code. Software pipelining was originally developed for hand microcoding vector operations. The technique has been used in the compilers for the ESL polycyclic machine [47], Cydrome's Cydra 5 [10], and the FPS-164 computer [56]. The problem of finding the optimal schedule has been shown to be NP-complete. The polycyclic and Cydra compilers depend on expensive specialized hardware support to simplify the scheduling problem; the FPS compiler uses scheduling heuristics, but limits the application of software pipelining to a restricted class of loops.

This work extends the previous results of software pipelining in two ways. First, we show that expensive hardware support is not necessary for software pipelining by improved scheduling heuristics and a new optimization called *modulo variable expansion*. The latter has some of the functionality of the specialized hardware proposed in the polycyclic machine and achieves similar performance. Empirical results indicate that the proposed scheduling heuristics often produce near-optimal schedules.

Second, software pipelining has previously been applied only to loops with straight-line code bodies. The hierarchical reduction scheme described in the next chapter makes software pipelining applicable to all loops. In particular, all innermost loops, including those containing conditional statements, can be software pipelined. If the number of iterations in the innermost loop is small, we can software pipeline the second level loop as well to obtain the full benefits of this technique.

Software pipelining, as addressed here, is the problem of scheduling operations within an iteration, such that the iterations can be pipelined to yield optimal throughput. The assumed machine model is that the machine contains one or more, possibly different, possibly pipelined, functional units. Software pipelining has also been studied under different contexts. The software pipelining algorithms proposed by Su et al. [52, 53], and Aiken and Nicolau [2], assume that the schedules for the iterations are given and cannot be changed. This constraint was relaxed in Aiken and Nicolau's subsequent paper in which they studied the problem of generating an optimal software pipelined schedule for a machine with infinitely many resources. Ebcioglu also studied a similar problem, and suggested an algorithm for software pipelining loops with conditional statements [14]. Finally, Weiss and Smith compared the results of using loop unrolling and software pipelining to generate scalar code for the Cray-1S architecture [57]. However, their software pipelining algorithm only overlaps the computation from at most two iterations. The unfavorable results obtained for software pipelining can be attributed to the particular algorithm rather than the software pipelining approach.

This chapter concentrates on software pipelining loops whose body consists of a single basic block. (Other cases are discussed in the next chapter). We first introduce the software pipelining technique, present the mathematical formulation of the scheduling problem, and describe the scheduling heuristics. We then introduce the modulo variable expansion optimization. The scheduling algorithm is derived from the techniques used for the FPS and the polycyclic machines; a detailed comparison between the techniques used is presented at the end of the chapter. In addition, this chapter includes a study of the specialized interconnection proposed for the polycyclic machine, evaluating the tradeoff between hardware and software complexity. Comparison with trace scheduling is presented in the next chapter after the description of hierarchical reduction.

5.1. Introduction to software pipelining

The effect of software pipelining can be illustrated by the following example:

```
FOR i := 1 TO 9 DO
BEGIN
    Receive (L, X, a);
    Send (R, X, a+b);
END;
```

Consider compiling this program into code for the Warp cell. Although the processor can support the initiation of one floating-point multiplication and one floating-point addition every instruction, the latency for each operation is long. At the machine instruction level, floating-point arithmetic operations have a seven-instruction, or seven-clock, latency; this latency includes a two-clock delay in fetching data from the multi-ported register file, and a five-clock floating-point operation. Using the same notation as in Chapter 3, the most compact instruction sequence possible for a single iteration of this loop is given in Figure 5-1. The schedule is sparse due to the heavy pipelining in the data path. If we simply iterate this schedule, the throughput of the loop is only 1 iteration

```
RecX.
Add.      ; a+b
.         ; 7-stage pipelined adder
.
.
.
.
SendX.    ; Send result from adder out
```
Figure 5-1: Object code for one iteration in example program

every 8 clock ticks, and no resources are used more than 1/8th of the time.

Trace scheduling relies on unrolling to improve the throughput of a loop [19]. The body of the loop is unwound some number of times and code compaction is performed on the unrolled source code. Unfortunately, there is no clear criterion to determine the suitable degree of unrolling. The utilization almost always improves as more iterations are unrolled; however, the problem size and the resulting code size increase likewise. Suppose the loop body of the example is unrolled 8 times, the optimal schedule of the body of the unrolled loop is shown in Figure 5-2.

```
L:RecX.
   Add, RecX.
        Add, RecX.
             Add, RecX.
                  Add, RecX.
                       Add, RecX.
                            Add, RecX.
   SendX,                       Add, RecX.
        SendX,                       Add.
             SendX.
                  SendX.
                       SendX.
                            SendX.
                                 SendX.
   CJump L,                           SendX.
```
Figure 5-2: Optimal schedule for 8 iterations

This microcode sequence assumes that the number of iterations is divisible by 8. Each row in the figure corresponds to operations in an instruction, and each column corresponds to the computation of one iteration of the loop in the source program. The operation **CJump L** branches back to label **L** if there are more iterations to execute. In Warp, one branch operation can execute in parallel with other micro-operations. Unrolling the loop 8 times improves the throughput to 8 iterations every 15 clocks. For this program, unrolling the loop k times increases the utilization of the resources to $k/k+7$.

In software pipelining, we do not wait till the completion of an iteration before initiating the next iteration. The iterations in the loop are executed in a pipelined fashion, with multiple iterations executing at the same time in the steady state. Let us consider the same example again. It is obvious from Figure 5-2 that the rate of initiating one iteration every instruction can be kept up until we run out of iterations. Intuitively, the eight instructions in an iteration of the loop can be viewed as an 8-stage pipeline that accepts a new iteration every clock (Figure 5-3). In the

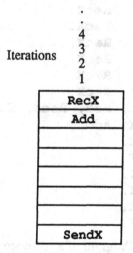

Figure 5-3: The software pipeline

steady state, 8 consecutive iterations of the loop are executed in parallel, with each one in a different stage of its processing. It is this analogy to a hardware pipeline that software pipelining derives its name.

The schedule for the software pipelined loop can be succinctly represented by a program that is only about twice as long as the program for a single iteration (Figure 5-4). The program in the figure assumes that there are at least eight iterations in the loop. Instructions 1 to 7 are called the *prolog*, in which more and more iterations of the loop start executing. The *steady state* is reached in clock 7, and is repeated until all iterations have been initiated. In the steady state, eight iterations are in progress at the same time, with one iteration starting up and one finishing off every instruction executed. On leaving the steady state, the iterations currently in progress are completed in the *epilog*, instructions 9 to 15. This program achieves the *optimal* computation time by completing n iterations in $n+7$ clock ticks, where n is the number of iterations in the loop. After a latency of eight clock ticks, the iterations are executed at the optimal throughput of one iteration every clock.

```
        RecX,
        Add,    RecX.
        Add,    RecX.
        Add,    RecX.
        Add,    RecX.
        Add,    RecX.
        Add,    RecX.
L:      SendX,  Add,  RecX,  CJump L.
        SendX,  Add.
        SendX.
        SendX.
        SendX.
        SendX.
        SendX.
        SendX.
```

Figure 5-4: Program of a software pipelined loop

Software pipelining is unique in that the pipeline stages in the functional units in the data path are not emptied at iteration boundaries; the pipelines are filled and drained only on entering and exiting the loop. The significance of software pipelining are: (1) optimal throughput is achievable, and (2) the code generated is extremely compact.

5.2. The scheduling problem

Software pipelining was originally introduced for scheduling hardware pipelines, and the problem was formulated as inserting delays between hardware units to increase the overall throughput of the system [44]. New input is accepted by the hardware pipeline at regular periodic intervals. The software analog of this approach is to derive a same schedule for each iteration, then initiate and stagger the iterations at periodic intervals. The repeating sequence of states are captured by a loop of microinstructions, one for each state. To minimize the number of states, and thus the size of the microcode, Rau and Glaeser [47] suggested keeping the initiation intervals between every pair of iterations the same. Therefore, to generate the object code for a software pipelined loop, we need the schedule of an iteration and the *iteration initiation interval* [47], the interval at which the iterations of the loop are initiated. In the example above, the schedule of an iteration is given in Figure 5-1, and the initiation interval is one.

The objective of software pipelining is to find a schedule for an iteration that can be initiated at shortest intervals. The following describes the scheduling constraints of the problem and presents the mathematical formulation of the problem.

5.2.1. Scheduling constraints

As in the scheduling problem for basic blocks, there are two kinds of scheduling constraints: resource and precedence constraints. The difference, however, is that both sets of constraints are defined in terms of the initiation interval of the schedule.

Resource constraints. If iterations in a software pipelined loop are initiated every sth clocks, then every sth instruction in the schedule of an iteration is executed simultaneously, one from a different iteration. Therefore, the total resource requirement of every sth instruction cannot exceed the available limit. Using the terminology defined in Chapter 4, we represent the resources required clock i by a *modulo resource usage function,* $\bar{\rho}^s$:

$$\bar{\rho}^s(i)= \sum^{k\in Z} \rho(i+ks).$$

The resource constraint is thus:

$$\forall\ 0\leq i < s,\ \ \sum^{v\in V} \bar{\rho}^s_v(i - \sigma(v)\bmod s) \leq R.$$

The overlapped resource usage from different iterations is illustrated in Figure 5-5(a). The resource usage for each iteration is represented by

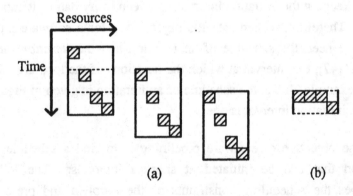

(a)   (b)

Figure 5-5: (a) Overlapped resource usage and
(b) modulo resource usage function

a reservation table. The entry in column i and row j represents the use of the ith resource in step j. A shaded entry corresponds to the use of the corresponding resource. (In this representation, the machine model can have only one unit of each kind of resources. This is only a limitation to simplify the illustration; the mathematical formulation of the problem and the proposed algorithm support the presence of multiple units of identical functional units.) Figure 5-5(b) shows the resource usage in the steady state. This modulo resource usage table can be derived by folding the resource usage of each iteration into a table of size *s*, the initiation interval.

Precedence constraints. As iterations of a loop are overlapped in software pipelining, global data dependencies between operations from different iterations must be considered. Consider the following example:

```
FOR i := 1 to 100 DO
BEGIN
    a := a + 1.0;
END;
```

The minimum delays between the read and write operations of the variable **a**, assuming the machine characteristics of the Warp cell, are depicted in Figure 5-6(a). Due to the pipelined addition, the write opera-

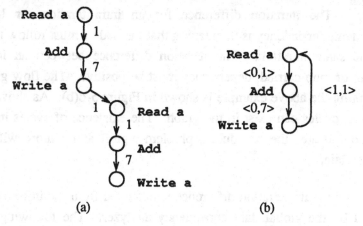

Figure 5-6: (a) Delays between operations from two iterations, and (b) flow graph representation

tion must be scheduled at least 8 clocks after the read operation of the current iteration. The read operation, however, must wait for the write operation from the previous iteration to complete. (Please note that here we explain the algorithm as if the nodes are scheduled in the order they are executed. In reality, the graph is inverted, and operations that must execute first are scheduled last, as explained in Section 4.3.3 of Chapter 4. As this reversal in scheduling order is orthogonal to the software pipelining technique, we ignore this reversal in the explanation here to make it more intuitive.)

The precedence constraints are captured by labeling the edges not with a single delay as in the basic block scheduling problem, but by a tuple: the minimum number of iterations separating the related instances of the nodes and the minimum delay between them. When we say that the minimum iteration difference on an edge (u,v) is p and the delay is d, it means that the node v must execute d or more clocks after node u from the p th previous iteration. That is,

$$\sigma(v)-(\sigma(u)-s \cdot p(u,v)) \geq d(u,v), \quad \text{or} \quad \sigma(v)-\sigma(u) \geq d(u,v)-s \cdot p(u,v),$$

where s is the initiation interval. Since a node cannot depend on a value from a future iteration, the minimum iteration difference is always non-negative. The iteration difference for an intra-iteration, or loop-independent, dependency is 0, meaning that the node v must follow node u in the same iteration. The iteration difference between an inter-iteration, or loop-carried, dependency must be positive. The flow graph representing the above example is shown in Figure 5-6(b). As shown in the figure, cycles can exist in the graph. The existence of cycles in the graph complicates the scheduling problem significantly; more will be discussed later.

The minimum iteration difference is extracted from the information obtained by the global data dependency analyzer. The following example illustrates the different kinds of dependency:

```
FOR i := 1 to 100 DO
BEGIN
        a[i] := a[i] + 1.0;
        c[i] := c[i-2] + 1.0;
        d[i] := d[3] + 1.0;
END
```

The assignments to variable a[i] in different iterations are independent, so there is no inter-iteration precedence constraint between these operations. The variable c[i-2] refers to the value stored into the array two iterations ago; therefore, the iteration difference is 2. And in the case of the variable d, the iteration difference between the related instances of the nodes is not constant. The worst case is assumed and we label the precedence constraint with the minimum separation of one iteration.

5.2.2. Definition and complexity of problem

The definition of the software pipelining problem is as follows:

Let R be the resource configuration vector, where $R(i)$ is the number of units of resource i available in the system configuration. A resource usage vector r is a tuple of the same length where $0 \leq r(i) \leq R(i)$. Let $G=(V,E)$ be a directed graph. Associated with each vertex v are the following attributes:
- length $l(v)$, and
- a resource usage function ρ_v, where

$$\rho_v(i)=\begin{cases} \text{resource usage vector for step } i, & 0 \leq i < l(v) \\ <0,0,\ldots,0> & \text{otherwise} \end{cases}$$

The number of jth resource used in step i is denoted by $\rho_v(i,j)$.

Associated with each directed edge e are two quantities:
- the minimum iteration difference by which the related nodes are separated: $p(e)$, and
- the minimum delay between the nodes, $d(e)$.

The problem is to find the minimum initiation interval s and a schedule $\sigma: V \to N$ such that the precedence and resource constraints are satisfied, i.e.,

$$\forall e=(u,v) \in E, \quad \sigma(v)-\sigma(u) \geq d(e)-s \cdot p(e)$$

and

$$\forall 0 \leq i < s, \quad \sum_{v \in V} \overline{\rho}_v^s(i - \sigma(v) \bmod s) \leq R.$$

The problem of whether a schedule can be found for a given initiation interval is NP-complete; this can be shown by reduction from the resource constrained scheduling problem [27]. In the resource constrained scheduling problem, the tasks do not have precedence constraints and they all take unit execution time. To ask if a schedule can be found in n clocks can be reduced to asking if a schedule can be software pipelined with an initiation interval n.

5.3. Scheduling algorithm

There have been two approaches in response to the complexity of this problem: (1) change the architecture, and thus the characteristics of the constraints, so that the problem is no longer NP-complete, and (2) use software heuristics. The first approach is used in the polycyclic architecture; a specialized crossbar is used to make optimizing a subset of loops tractable. The second approach is used in the FPS compiler. We also use software heuristics; we have improved the heuristics and extended the applicability of the technique to include all loops. Comparison of our algorithm with different techniques will be presented after the description of the algorithm.

The software pipelining algorithm is complicated because of two reasons. First, the scheduling constraints are defined in terms of the initiation interval, and this makes finding an approximate solution to this NP-complete problem difficult. Since computing the minimum initiation interval is NP-complete, a standard approach is to first schedule the code using heuristics, and then determine the initiation interval permitted by the schedule. However, since the scheduling constraints are a function of the initiation interval, if the initiation interval is not known at scheduling time, the schedule produced is unlikely to permit a good initiation interval.

To resolve this circularity, the FPS compiler uses an iterative approach: first establish a lower and an upper bound on the initiation interval, then use binary search to find the smallest initiation interval for which a schedule can be found [56]. (The length of a locally compacted iteration can serve as an upper bound; the calculation of a lower bound is described below). We also use an iterative approach, but we use linear instead of binary search. That is, we try to find a schedule using the lower bound of the initiation interval as the target interval. We iterate this process if we fail to find such a schedule by increasing the target initiation interval by one clock tick at a time. The rationale is as follows: Although the probability that a schedule can be found generally increases with the value of the initiation interval, schedulability is not monotonic, as explained below. Especially since empirical results show that, in the case of Warp, a schedule meeting the lower bound can often be found, sequential search is preferred.

The second cause for the complexity of the scheduling algorithm is the presence of cycles in the graph. For a given initiation interval, an *acyclic* graph can be scheduled using the list scheduling algorithm described in the previous chapter, substituting the resource function with the modulo resource usage function. Cycles in the flow graph, however, make designing an effective non-backtracking scheduling algorithm difficult. The algorithm proposed here first finds the *strongly connected components* of the graph, schedules the individual strongly connected components using a different set of heuristics, and finally schedules the components themselves using the list scheduling algorithm for acyclic graphs.

In the following, we discuss in detail the bounds on the initiation interval, the algorithm for acyclic graphs, and finally, the algorithm for cyclic graphs.

5.3.1. Bounds on the initiation interval

A lower bound on the initiation interval can be derived from the resource constraints for all graphs; if the graph is cyclic, a bound is further imposed on the initiation interval by the precedence constraints. As shown in Chapter 7, empirical results show that the bound obtained using the formulae below is often strict.

Resource constraints. If an iteration is initiated every s clocks, then the total number of resource units available in s clocks must at least cover the resource requirement of one iteration. Therefore, the bound on the initiation interval due to resource considerations is the maximum of the total number of times each resource is used divided by the available units per clock:

$$S_R = \max_k \left\lceil \frac{\sum^{v \in V, 0 \leq i < l(v)} \rho_v(i,k)}{R(k)} \right\rceil .$$

To see that the bound is not necessarily tight even if there are no precedence constraints between the nodes, consider a loop whose body consists of two nodes with resource functions:

Resource conflicts preclude such a loop from being pipelined with initiation interval S_R (=2).

Precedence constraints. Cycles in precedence constraints impose a minimum distance between operations from different iterations. The initiation interval must be large enough for such delays to be observed. For example, the precedence constraints in Figure 5-6 impose a delay of 8 clock ticks between read operations from consecutive iterations. That is, the initiation interval has to be no smaller than 8 clocks. We define the minimum delay and minimum iteration difference of a path to be the sum of the minimum delays and minimum iteration differences of the edges in the path, respectively. Let s be the initiation interval, and c be a cycle in the graph. Since

$$\sigma(v)-\sigma(u)\geq d(e)-s\cdot p(e)$$

we get:

$$d(c)-s\cdot p(c)\leq 0.$$

We note that if $p(c)=0$, then $d(c)$ is necessarily less than 0 by the definition of a legal computation. Therefore, the bound on the initiation interval due to precedence considerations is

$$S_E=\max_c \left\lceil \frac{d(c)}{p(c)} \right\rceil, \quad \forall \text{ cycle } c \text{ whose } p(c)\neq 0.$$

The lower bound of the initiation interval is given by the maximum of the lower bounds established by resource and precedence considerations.

We can also bound the initiation interval *from above* by list scheduling one iteration of the loop. The length of the schedule serves as an upper bound of the initiation interval. Unless a schedule can be found for an initiation interval shorter than the schedule of an iteration, there is no advantage to software pipelining. We call this upper bound S_{max}.

The probability of finding a schedule generally increases with the value of the initiation interval. However, schedulability is not monotonic. For example, a node with the following resource usage can be initiated every 2, but not 3, clocks.

Especially since empirical results reported in Chapter 7 show that the lower bound can almost always be met, sequential search is more preferable than binary search.

5.3.2. Scheduling an acyclic graph

For acyclic graphs, the initiation interval s is bounded from below by only S_R. We try to find a software pipeline schedule starting with this lower bound. The algorithm we use to schedule an acyclic graph for a target initiation interval is the same as that used in the FPS compiler, which itself is derived from the list scheduling algorithm used in basic block scheduling [17]. The difference is that the modulo resource usage function is used in computing resource conflicts. By the definition of modulo resource usage, if we cannot initiate a node in s consecutive clock ticks due to resource conflicts, it will not fit in the current schedule. When this happens, we abort the attempt to find a schedule for the given initiation interval and repeat the scheduling process with a greater interval value. If no schedule is found for an initiation interval value less than S_{max}, simple list scheduling is used.

An example of an acyclic flow graph is shown in Figure 5-7(a).

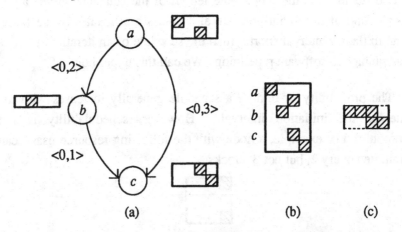

(a)   (b)   (c)

Figure 5-7: Scheduling an acyclic graph: (a) an example flow graph,
(b) resource usage of an iteration, and
(c) modulo resource usage

Since the third resource is used twice in an iteration, the initiation interval is at least two clocks. We first attempt to schedule the loop for a

target initiation interval of two clock ticks. By following the topologically ordering of the precedence constraints, we schedule node a in clock
0, then node b in clock 2. Although precedence constraints permit node
c to be scheduled in clock 3, execution of the node is delayed until clock
4 to avoid creating a conflict in the modulo resource reservation table.
The resource usage of a single iteration and the modulo resource reservation table are shown in Figure 5-7(b) and (c), respectively. The rows
in the reservation table are labeled with the nodes that are initiated in the
corresponding clock ticks. This example shows that a locally optimal
schedule for a single iteration does not necessarily result in a globally
optimal pipelined schedule for the loop.

The algorithm for software pipelining an acyclic graph is given in
Figure 5-8. The function *ListSchedule* is now modified from the definition in Chapter 4 to find a schedule for a basic graph within a loop, for a
given initiation interval s. Scheduling a basic block not nested any loops
is a special case. This can be achieved by specifying the initiation interval to be infinity. The function *SatisfyResourceConstraints* searches for
a slot to initiate the node between the given lower and upper bounds. If
no such slot is found, the function reports that the attempt to schedule the
node is unsuccessful. *ListSchedule* is called iteratively with increasing
values of target initiation intervals until a schedule is found or until the
upper bound is met.

FUNCTION *SchedAcyclicGraph* (V, E)
BEGIN

$$S_R := \max_k \left\lceil \frac{\displaystyle\sum^{v \in V, 0 \le i < l(v)} \rho_v(i,k)}{R(k)} \right\rceil ;$$

$S_{max} :=$ length of *ListSchedule* (V, E, ∞);
FOR $s := S_R$ TO S_{max} DO
BEGIN
    $\sigma := $ *ListSchedule* (V, E, s);
    IF $\sigma \ne \perp$ THEN RETURN $(<\sigma, s>)$;
END;

$\quad$ RETURN ($<ListSchedule$ $(V, E, \infty), S_{max}>$);
END;

FUNCTION $ListSchedule$ (V, E, s)
BEGIN
$\quad$ $Ready$:= root nodes of V;
$\quad$ $Sched$:= $\emptyset$;
$\quad$ WHILE $Ready \neq \emptyset$ DO
$\quad$ BEGIN
$\quad\quad$ v := highest priority node in $Ready$;
$\quad\quad$ Lb := $SatisfyPrecedenceConstraints$ $(v, Sched, \sigma, s)$;
$\quad\quad$ $\sigma(v)$:= $SatisfyResourceConstraints$ $(v, Sched, \sigma, s, Lb, Lb+s-1)$;
$\quad\quad$ IF $\sigma(v) = \perp$ THEN RETURN($\perp$);
$\quad\quad$ $Sched$:= $Sched \cup \{v\}$;
$\quad\quad$ $Ready$:= $Ready-\{v\}+\{u | u \notin Sched \wedge \forall (w,u) \in E, w \in Sched\}$;
$\quad$ END;
$\quad$ RETURN(σ);
END;

FUNCTION $SatisfyPrecedenceConstraint$ $(v, Sched, \sigma, s)$
BEGIN
$\quad$ RETURN ($\max_{u \in Sched} \sigma(u)+d(u,v)-s \cdot p(u,v)$)
END

FUNCTION $SatisfyResourceConstraints$ $(v, Sched, \sigma, s, Lb, Ub)$
BEGIN
$\quad$ FOR i := Lb TO Ub DO
$$\text{IF } \forall 0 \leq j < s, \overline{\rho}_v^s(j-i \bmod s)+ \sum^{u \in Sched} \overline{\rho}_u^s(i+j-\sigma(u) \bmod s) \leq R$$
$\quad\quad$ THEN RETURN(i);
$\quad$ RETURN($\perp$);
END;

Figure 5-8: A software pipelining algorithm for acyclic graphs

5.3.3. Scheduling a cyclic graph

Here we first present the overview of the algorithm, the rationale and full details of the algorithms are described below. We preprocess the graph by finding the strongly connected components [54] in the graph, and solving the all-points longest path problem for each component [11, 25]. As the weight on the precedence constraints is expressed as a linear function of the initiation interval, the longest path algorithm is modified to operate on the symbolic value of the target initiation interval to avoid recomputing the longest paths for each initiation interval.

As in the case of acyclic graphs, the main scheduling step is iterative. For each target initiation interval, the strongly connected components are first scheduled individually. The original graph is then reduced by representing each strongly connected component as a single vertex. The resulting graph is acyclic, and the acyclic graph scheduling algorithm is used.

The scheduling algorithm for strongly connected components also uses the framework of list scheduling. The nodes are scheduled according to a topological ordering of *intra-iteration* precedence constraints. As each node is scheduled, we use the precomputed longest path information to update the range within which each remaining node must be scheduled. A node is scheduled in the earliest possible slot within the allowable range. This scheme takes advantage of the relaxation of the scheduling constraints as the initiation interval increases, as explained below.

5.3.3.1. Combining strongly connected components

In the discussion to follow, we shall use the simple cyclic graph in Figure 5-9(a) as an example. (Note that the edge from node b to c has a negative delay; negative delays arise, for example, when step i of a node depends on step j of another node, and $j<i$.) Although each machine resource is used only twice in each iteration of the loop, the lower bound of the initiation interval is three clock ticks, as determined by the cycle of edges $a \rightarrow b \rightarrow c \rightarrow a$. The nodes a, b and c, being nodes of a strongly connected component, are scheduled together. These nodes are then reduced to a single node e, as shown in Figure 5-9(b). The resulting graph is acyclic and the acyclic scheduling technique described in Figure 5-8 is applied.

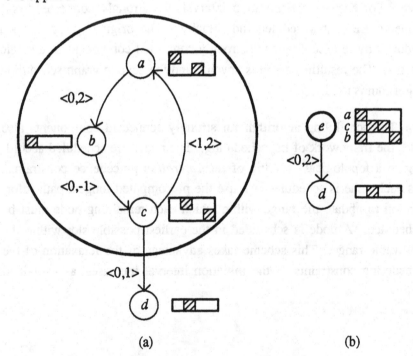

(a)   (b)

Figure 5-9: Scheduling a cyclic graph:
(a) original, and (b) reduced flow graph

In reducing the nodes in a strongly connected component into one node, we are combining the corresponding micro-operation sequences into one long micro-operation sequence. The ith step of a sequence scheduled to initiate in clock tick n is executed in step $n+i$ of the assembled sequence; resource and precedence constraints involving step i in the sequence must therefore be transferred to step $n+i$ of the assembled sequence. In the above example, the resource usage of node e reflects the schedule and the resource usage of nodes a, b and c. The precedence constraint between nodes c and d is mapped onto an edge between nodes e and d in the reduced graph.

The formal definition of a reduced graph is given below:

Let the nodes V in graph $G=(V,E)$ be partitioned into subsets V_i, where $\cup_i V_i=V$. Suppose the nodes in the subsets V_i have been scheduled with schedules $\{\sigma_{V_i}\}$, $G'=(\{u_i\},E')$ is a reduced graph, if and only if

1. $\rho_{u_i}(j)= \displaystyle\sum_{v \in V_i} \rho_v(j-\sigma_{V_i}(v))$.

2. For every edge
$$(v_1,v_2)\in E, \text{where } v_1 \in V_i, v_2 \in V_j \text{ and } V_i \neq V_j$$
   there is an edge
$$(u_i,u_j)\in E', \text{where } p\,(u_i,u_j)=p\,(v_1,v_2)$$
$$\text{and } d\,(u_i,u_j)=d\,(v_1,v_2)-\sigma_{V_i}(v_1)+\sigma_{V_j}(v_2).$$
   If multiple edges are defined between the same pair of nodes, then the maximum cost is used.

3. There are no more edges in E' other than those obtained from 2.

A reduced graph has the characteristic that some of the nodes in the graph may have rather heavy resource usage. In scheduling basic blocks, the resource usage is not used in prioritizing between ready nodes because it is insignificant compared to the long delays separating the nodes.

Here, nodes representing more than one node in the original graph are given a higher priority because of their greater resource requirements. The scheduling algorithm is otherwise the same.

5.3.3.2. Scheduling a strongly connected component

The same basic framework of list scheduling is used for scheduling a strongly connected component for a given initiation interval. The heuristics, however, must be revised. We cannot sort the nodes topologically; we cannot satisfy precedence constraints by merely considering those nodes that are already scheduled; and we cannot use the maximum height of the node as its priority. List scheduling is a non-backtracking algorithm and only one schedule is examined for each value of the initiation interval. We must ensure that the schedule examined has a reasonable chance of success. Experimentation with different schemes helped identify two desirable properties for the heuristics:

1. Partial schedules constructed at each point of the scheduling process should not violate any of the precedence constraints in the graph. In other words, were there no resource conflicts with the remaining nodes, each partial schedule should be a partial solution. Whenever such a partial schedule cannot be found, the process for the particular initiation interval should be considered unsuccessful and be terminated to reduce the compilation time.

2. The heuristics must be sensitive to the value of the initiation interval. An increased initiation interval value relaxes the scheduling constraints, and the scheduling algorithm must take advantage of this opportunity. It would be futile if the scheduling algorithm simply retries the same schedule that failed.

Both properties are exhibited by the heuristics in the scheduling algorithm for acyclic graphs, as well as the heuristics for strongly connected components, described below.

Satisfying precedence constraints. By the definition of strongly connected components, the entire set of precedence constraints in the component must be taken into consideration in scheduling each node. Our solution is to first precompute the maximum constraints between every pair of nodes by computing the closure of the constraints. This expensive computation is performed only once by representing the initiation interval with a symbolic value. After scheduling each node, we use this precomputed information to compute the *precedence constrained range* of the remaining unscheduled nodes. The precedence constrained range of a node for a given partial schedule is the legal range of clock ticks in which the node can be initiated without violating the precedence constraints of the graph. To ensure that all partial schedules satisfy all precedence constraints in the graph, a node is only allowed to be scheduled within its precedence constrained range.

The computation of the closure of precedence constraints is equivalent to the all-points longest path problem. Since all cycles in the graph have nonpositive costs, our all-points longest path problem is analogous to the all-points shortest path problem with no negative cycles. Dantzig et al. [11] showed that this problem can be solved using Floyd's algorithm [25].

In the closure computation, the cost of a path or an edge is represented as a function of the initiation interval s: $d(e)-s \cdot p(e)$. The precomputed result can be used for different values of s by simply substituting s with the particular initiation interval value. The cost of the longest path between two nodes is represented as a set of tuples $\{<p_1,d_1>, \ldots, <p_n,d_n>\}$ to capture all possible maximum costs between a pair of nodes for all values of s.

As we compute the longest path between every two points, we also find all the cycles in the graph. As discussed in Section 5.3.1, the lengths of the cycles bound the value of the initiation interval s. This bound can in turn be used to reduce the size of the maximum cost set of a path. Suppose we have a set $\{<0,4>,<1,6>\}$. Since

$$\max (<0,4>,<1,6>)=\begin{cases} 4, & s \geq 2 \\ 6-s, & \text{otherwise,} \end{cases}$$

if we know that s is greater than 2, we can simply drop $<1,6>$ out of the cost set.

In general, for a pair of costs $<p_1,d_1>$ and $<p_2,d_2>$, and a bound on the initiation interval, S, $<p_1,d_1>$ can be dropped from the cost set if

$$p_1 >= p_2 \quad \text{and} \quad d_1 - d_2 < (p_1 - p_2) \cdot S$$

The algorithm for finding the longest path and a lower bound on the initiation interval is given in Figure 5-10.

Figure 5-11 shows the closure computed using the algorithm described above for the strongly connected component shown in Figure 5-9.

Once the closure is computed, the precedence constrained range of a node can be calculated as follows: Let $Lb(u)$ and $Ub(u)$ be the current lower and upper bounds of the precedence constrained range of node u, respectively. Suppose node v is scheduled, and $<p_{vu},d_{vu}>$ and $<p_{uv},d_{uv}>$ are the maximum constraints from v to u, and v to u, respectively. Then,

$$Lb(u)=\max (Lb(u), \sigma(v)+d_{vu}-s \cdot p_{vu}), \text{and}$$
$$Ub(u)=\min (Ub(u), \sigma(v)-d_{uv}+s \cdot p_{uv}).$$

Figure 5-12(a) shows the precedence constrained ranges for nodes b and c after node a has been scheduled in time 0. The edges used in calculating the precedence constrained ranges are shown in the diagram. After scheduling node a, we have the choice to schedule b or c next. Figures 5-12(b) and (c) show the update of the remaining node for the different alternatives.

Ordering the nodes. The order in which the nodes are scheduled is crucial to the success of the algorithm. The important point is that the ordering must permit the scheduler to take advantage of the relaxed constraints as the target initiation interval is increased. Let us illustrate this

---

```
FUNCTION LongestPath (V, E, S)
BEGIN
     n := ‖ V ‖;
     FOR i := 1 TO n DO
          FOR j := 1 TO n DO
               C_{ij} := ∅;

     FOR (v_i, v_j) ∈ E DO
          C_{ij} := {<p(v_i,v_j), d(v_i,v_j)>};

     FOR k := 1 TO n DO
          FOR i := 1 TO n DO
               FOR j := 1 TO n DO
               BEGIN
                    C_{ij} := MaxCost(C_{ij}, AddCost(C_{ik}, C_{kj}), S)
```

$$S := \max(S, \max_{<p,d> \in AddCost(C_{ji}, C_{ij})} \left\lceil \frac{d}{p} \right\rceil);$$

```
               END;
          RETURN(<C, S>);
END;

FUNCTION AddCost (C_1, C_2)
BEGIN
     RETURN({<p_1+p_2, d_1+d_2> | <p_1, d_1> ∈ C_1 and <p_2, d_2> ∈ C_2});
END;

FUNCTION MaxCost (C_1, C_2, S)
BEGIN
     RETURN({<p, d> | <p, d> ∈ C_1 ∪ C_2 and ∄ <p_1, d_1> ∈ C_1 ∪ C_2,
                    where p >= p_1 and d - d_1 < (p - p_1)· S});
END;
```

Figure 5-10: Algorithm to find the longest paths and cycles
in a strongly connected component

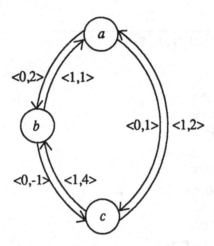

Figure 5-11: Closure of the strongly connected component in Figure 5-9

with an example. Suppose we use the deadline, the upper bound of the precedence constrained range, to prioritize the nodes. The node with the closest deadline is scheduled first. Given the graph in Figure 5-12(a), node c would be scheduled after node a because c has a closer deadline than b. Suppose there is no resource conflict between node a and c, node c is scheduled to execute in time 1, as early as the precedence constraints permit. Unfortunately, scheduling c at time 1 limits the schedulable range of node b to exactly one time slot, time 2. This value remains unchanged regardless of the value of the initiation interval s. Thus if scheduling node b in clock 2 happens to cause a resource conflict, it is futile to reiterate the scheduling process with a greater value of s.

On the other hand, if b is scheduled next instead of c, this problem does not exist. Suppose the node b is scheduled in time 2, the schedulable range of c will then be between time 1 and $s-1$. This range increases with the value of s. Therefore, if we cannot schedule node c for a given initiation interval, the chance of success improves with a larger value of s in each reiteration of the scheduling process.

Thus we propose the following algorithm: we schedule nodes in a topological ordering of the *intra-iteration subgraph*. The vertex set of

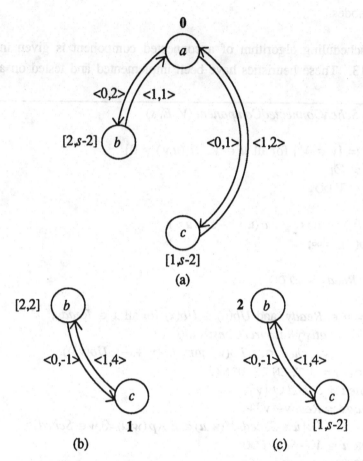

Figure 5-12: Examples of precedence constrained ranges

the intra-iteration subgraph is the same as the original graph, but the edge set contains only those edges with a zero minimum iteration difference, $p=0$. As these edges only relate operations within the same iteration, the intra-iteration subgraph must be acyclic. With this ordering, the *upper* bound on the schedulable ranges of all unscheduled nodes is always a function of the initiation interval. As the initiation interval increases, so does the chance of success. The scheduling problem approaches that of an acyclic graph as the value of the initiation interval increases. The

deadline of the precedence constrained range is only used to prioritize
the ready nodes.

The scheduling algorithm of a connected component is given in
Figure 5-13. These heuristics have been implemented and tested on a

---

FUNCTION *SchedConnectedComponent* (V, E, s)
BEGIN
    Ready := $\{v \in V \mid$ for all $u \in V, \; p(u,v) \neq 0\}$
    Sched := $\varnothing$;
    FOR $v \in V$ DO
    BEGIN
        $Lb(v) := \max_{u \in V} d(u,v) - s \cdot p(u,v)$;
        $Ub(v) := \infty$;
    END;
    WHILE *Ready* $\neq \varnothing$ DO
    BEGIN
        $v := u \in$ *Ready*, and $Ub(u) \leq Ub(x)$ for all $x \in$ *Ready*;
        $\sigma(v) :=SatisfyResourceConstraints$
                $(v, Sched, \sigma, s, Lb(v), \min(Lb(v)+s-1, Ub(v)))$;
        IF $\sigma(v) = \perp$ THEN RETURN $(\perp)$;
        Sched := *Sched* $\cup \{v\}$;
        Ready := *Ready*$-\{v\}+$
                $\{u \mid u \notin Sched \wedge \forall(w,u) \in E \wedge p(w,u)=0, w \in Sched\}$;
        FOR $u \in V - Sched$ DO
        BEGIN
            $Lb(u) := \max(Lb(u), \sigma(v) + d(v,u) - s \cdot p(v,u))$;
            $Ub(u) := \min(Ub(u), \sigma(v) - d(u,v) + s \cdot p(u,v))$;
        END;
    END;
    RETURN (σ);
END;

Figure 5-13: Scheduling a strongly connected component

large set of programs; the results are reported in Chapter 7. There may
be many other heuristics that can produce similar results. The important

point, however, is to make sure that the scheduling heuristics have the two properties present in our algorithms: partial schedules constructed during the scheduling process should satisfy all precedence constraints in the graph, and the scheduling order must take advantage of the relaxation in constraints offered by increased values of the initiation interval.

5.3.3.3. Complete algorithm

Figure 5-14 contains the general procedure for software pipelining.

FUNCTION *SoftwarePipeline* (V, E)
BEGIN

$$S := \max_k \left\lceil \frac{\sum^{v \in V, 0 \le i < l(v)} \rho_v(i,k)}{R(k)} \right\rceil ;$$

S_{max} := computation time of a list schedule of the body of the loop;

$\{(V_i, E_i)\}$:= { connected components of (V, E)};

    FOR each (V_i, E_i) DO
        $<E_i, S> := LongestPath (V_i, E_i, S)$;

    FOR $s := S$ TO S_{max} DO
    BEGIN
        IF $\forall (V_i, E_i), SchedConnectedComponent (V_i, E_i, s) \ne \perp$ THEN
        BEGIN
            $(V', E') :=$ reduced graph of (V, E);
            $\sigma := ListSchedule (V', E', s)$;
            IF $\sigma \ne \perp$ THEN RETURN $(<\sigma, s>)$;
        END;
    END;
    RETURN $(<\perp, \perp>)$;
END;

Figure 5-14: Software pipelining

We first compute a lower bound on the initiation interval as imposed by resource constraints, and an upper bound by list scheduling one iteration of the loop. We next find the strongly connected components of the graph and compute the closure of the precedence constraints. We are then ready to find the initiation interval and schedule for the loop. We iterate within the bounds of the initiation interval; for each interval, we first try to schedule the strongly connected components individually, then schedule the reduced graph derived from the individual schedules. If any of the scheduling steps fails, we repeat the scheduling process with a greater initiation interval. Acyclic graphs can be handled by the same algorithm as each node is simply a trivial strongly connected component.

5.4. Modulo variable expansion

Modulo variable expansion is an optimization to remove unnecessary precedence constraints due to the reuse of a scalar variable in different iterations. The basic idea can be illustrated by the following code fragment in the body of a loop. A value is written into a register and used two clocks later:

```
Def(R1)
op
Use(R1)
```

If the same register is used by all iterations, then the write operation of an iteration cannot execute before the read operation in the preceding iteration. Therefore, the optimal throughput is limited to one iteration every two clocks. This code can be sped up by using different registers in alternating iterations:

```
    Def(R1),
    op,        Def(R2).
L:Use(R1),op,        Def(R1).
           Use(R2),op,        Def(R2),CJump L.
                    Use(R1),op.
                            Use(R2).
```

We call this optimization of allocating multiple registers to a variable in the loop *modulo variable expansion*. This optimization is a variation of the variable expansion technique used in vectorizing compilers [33]. The variable expansion transformation identifies those variables that are redefined at the beginning of every iteration of a loop, and expands the variable into a higher dimension variable, so that each iteration can refer to a different location. (Results of partial expressions also fall into the category of scalar variables that can be expanded). Consequently, the use of the variable in different iterations is thus independent, and the loop can be vectorized. Modulo variable expansion takes advantage of the flexibility of VLIW machines in scalar computation, and reduces the number of locations allocated to a variable by reusing the same location in non-overlapping iterations. The small set of values can even reside in register files, cutting down on both the memory traffic and the latency of the computation.

Without modulo variable expansion, the length of the steady state of a pipelined loop is simply the initiation interval. When modulo variable expansion is applied, code sequences for consecutive iterations differ in the registers used, thus lengthening the steady state. If there are n repeating code sequences, the steady state needs to be *unrolled n* times.

The algorithm of modulo variable expansion is as follows. First, we identify those variables that are redefined at the beginning of every iteration. Next, we pretend that every iteration of the loop has a dedicated register location for each qualified variable, and remove all *inter-iteration* precedence constraints between operations on these variables. Scheduling then proceeds as normal. The resulting schedule is then used to determine the actual number of registers that must be allocated to each variable. The lifetime of a register variable is defined as the duration between the first assignment into the variable and its last use. If the lifetime of a variable is l, and an iteration is initiated every s clocks, then at least $\lceil \frac{l}{s} \rceil$ number of values must be kept alive concurrently, in that many locations.

If each variable v_i is allocated its minimum number of locations, q_i, the degree of unrolling is given by the lowest common multiple of $\{q_i\}$. Even for small values of q_i, the least common multiple can be quite large and can lead to an intolerable increase in code size. The code size can be reduced by trading off register space. We observe that the minimum degree of unrolling, u, to implement the same schedule is simply $\max_i q_i$. This minimum degree of unrolling can be achieved by setting the number of registers allocated to variable v_i to be the smallest factor of u that is no smaller than q_i, i.e.,

$\min n$, where $n \geq q_i$ and $u \bmod n = 0$.

The increase in register usage is much more tolerable than the increase in code size of the first scheme for a machine like Warp.

Since we cannot determine the number of registers allocated to each variable until all uses of registers have been scheduled, we cannot determine if the register requirement of a partial schedule can be satisfied. Moreover, once given a schedule, it is very difficult to reduce its register requirement. Indivisible micro-operation sequences make it hard to insert code in a software pipelined loop to spill excess register data into memory.

In practice, we can assume that the target machine has a large number of registers; otherwise, the resulting data memory bottleneck would render the use of any global compaction techniques meaningless. The Warp machine has two 31-word register files for the floating-point units, and one 64-word register for the ALU. Empirical results reported in Chapter 7 show that they are large enough for almost all the user programs developed. Register shortage is a problem for a small fraction of the programs; however, these programs invariably have loops that contain a large number of independent operations per iteration. In other words, these programs are amenable to other simpler scheduling techniques that only exploit parallelism within an iteration. Thus, when register allocation becomes a problem, software pipelining is not as crucial. The best approach is therefore to use software pipelining aggressively, by assuming that there are enough registers. When we run out of registers, we then resort to simple techniques that serialize the execution of loop iterations. Simpler scheduling techniques are more amenable to register spilling techniques.

5.5. Code size requirement

Software pipelining produces highly efficient code without greatly increasing the object code size. Figure 5-15 shows the code of a loop whose steady state has been unrolled twice. The boxes of bold lines depict segments of code containing operation sequences from different iterations. The middle section of the block in the center represents the steady state. Iterations alternatively use the ''a'' and ''b'' sets of registers, and are labeled as such. The small circle marks the instruction that decides if the end of the loop is reached, and the dashed lines represent possible branches in the flow of control.

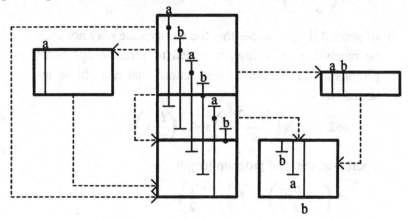

Figure 5-15: Code size requirement of software pipelining

The code length of a software pipelined loop is determined by the length of the schedule of an iteration (l), the length of code to evaluate the loop termination condition (c), the initiation interval (s) and the degree of unrolling (u). It also depends on the information available on the number of iterations in the loop. There are three cases:

1. If the number of iterations is known at compile time, we only have to generate one prolog, one steady state, and one epilog as shown in the center of the figure. The length of the loop is:

$$c+\left(\left\lceil\frac{l-c}{s}\right\rceil-1\right)s+us+(l-c-s) < l+us+l-s < 3l,$$

$$\text{since } u \le \left\lceil\frac{l}{s}\right\rceil.$$

2. If the loop is unrolled and the loop bound is not a compile-time constant, then multiple exits out of the steady state are needed, and we need $u-1$ more epilogs. This increases the length of the code by $(u-1)(l-c-s)$ instructions; the total length is therefore:

$$c+\left(\left\lceil\frac{l-c}{s}\right\rceil-1\right)s+us+u(l-c-s) < l+us+u(l-s) < (u+1)l$$

$$< \left(\left\lceil\frac{l}{s}\right\rceil+1\right)l.$$

3. In general, it is possible that the steady state may not even be reached; in this case, we need to generate additional prologs. The number of instructions that need to be inserted is:

$$s+2s+\ldots+\left(\left\lceil\frac{l-c}{s}\right\rceil-2\right)s < \frac{1}{2}\left(\frac{l}{s}-1\right)l.$$

Therefore, the total program length

$$< \left(u+\frac{l}{2s}+\frac{1}{2}\right)l < \left(\frac{3}{2}\left\lceil\frac{l}{s}\right\rceil+\frac{1}{2}\right)l.$$

The value $\left\lceil\frac{l}{s}\right\rceil$ is the number of iterations executed in parallel in the steady state, or the *degree of pipelining* in the software pipelined loop. The degree of software pipelining is limited to the number of concurrent operations that can be executed by the processor. The degree of pipelining is a close approximation to the speed up achieved by the software pipelining technique. Therefore the increase in code size is proportional to the speed up obtained. This code size increase is even lower if more information is available on the number of iterations. Especially in the case of compile-time loop bounds, the code size is at most three times that of one single iteration.

Moreover, the degree of pipelining tends to be inversely related to the number of operations in the loop. It is less important to find parallelism across multiple iterations when a single iteration contains a large number of operations. This is ideal because when the degree of pipelining is high, the length of an iteration tends to be short, and we can better afford a greater factor of increase in code size.

A more important metric than the code size for the entire loop is perhaps the length of the steady state. A program is typically much smaller than the data processed, therefore the overall size of the program is not important. However, for machines with instruction caches, it is important that the innermost loop fits entirely in the cache. In other words, the size of the steady state is crucial, but not the length of the prolog and epilog. Since the steady state is only us instructions long, the loop is typically much shorter than the code sequence for an unpipelined iteration. Also, with the global scheduling technique described in the next chapter, the prolog and epilog can be overlapped with operations outside the loop.

If the target machine architecture cannot support the increase in code size called for in the above scheme, FOR loops can be implemented with the following alternate scheme. We handle those cases that make multiple prologs and epilogs necessary by a separate loop. The compiler generates two loops, one software pipelined and the other not. The pipelined loop contains only one prolog, a steady state, and one epilog, and the other loop is produced by simply compacting one iteration of the loop body. Let n be the number of iterations to be executed, and k be the number of iterations started in the prolog of the pipelined loop. If $n < k$, then only the unpipelined loop is executed for n iterations. Otherwise, the unpipelined loop is executed for $n-k \bmod u$ iterations and the remaining iterations are executed on the pipelined loop. In this way, only one prolog and one epilog are needed, and the code length is at most four times the code for one iteration. The disadvantage, however, is that the prolog and epilog cannot be executed concurrently with other operations outside the loop.

5.6. Comparison with previous work

The following compares our software pipelining technique with that used in the compilers for the ESL polycyclic machine [47] and the FPS-164 computer [56].

5.6.1. The FPS compiler

Like Warp, the FPS-164 does not contain any specialized hardware to support software pipelining, and the scheduling algorithm relies on software heuristics. The algorithm for scheduling acyclic graphs presented here is similar to the algorithm used in the FPS compiler. The only difference is that the space of the initiation interval is searched sequentially in ours, whereas binary search is used in the FPS compiler. As discussed earlier, sequential search is used because schedulability is not monotonic with the initiation interval; moreover, empirical results show that a schedule can often be found with an initiation interval close to the lower bound. The major difference between the two approaches lies in the manner the cyclic graphs are handled. Also, we use modulo variable expansion to increase parallelism in the code.

The FPS compiler only software pipelines loops consisting of a single Fortran statement. Therefore, at most one inter-iteration data dependency relationship can be present in the flow graph. Although the scheduling algorithm for cyclic graphs used in the FPS compiler suffices for this restricted set of loops, it is unlikely to be effective for general cyclic graphs.

The FPS compiler handles cyclic graphs by introducing the concept of *EarlyStartTime* into the acyclic scheduling algorithm. The *EarlyStartTime* of a node can be preset before a scheduling attempt to prevent a node from being scheduled too early. Nodes are scheduled in a topological ordering of the intra-iteration subgraph. Since only precedence constraints with nodes that have already been scheduled are considered when scheduling each node, partial schedules constructed during the scheduling process do not necessarily satisfy all precedence constraints in the graph. If a node is discovered not to be schedulable because its precedence constraints are violated by the partial schedule, the culprit node is identified and its *EarlyStartTime* is modified to prevent it from being scheduled too early in the next attempt. This scheduling process is then repeated for the *same* initiation interval. While the algorithm suffices for graphs with a single inter-iteration data dependency relationship, this approach may require many attempts per initiation interval in the case of a general cyclic graph.

In our algorithm, the closure of precedence constraints is precomputed once before the first attempt of scheduling. This closure is used to update the precedence constrained range of each node cheaply in the scheduling process. The partial schedules always satisfy the precedence constraints of the graph and only one scheduling attempt is necessary for each initiation interval.

5.6.2. The polycyclic machine

The compiler for the polycyclic architecture relies on specialized hardware support. All functional units of a polycyclic machine are interconnected through a crossbar. This crossbar has storage at every crosspoint to serve as a dedicated buffer for each pair of functional units. Therefore, there is never any contention in reading or writing data by any pair of functional units. Each node in the flow graph thus consumes only one explicitly scheduled resource. This is significant because the scheduling problem for acyclic graphs now becomes tractable [44, 47]. The minimum initiation interval can be predetermined, and an optimal schedule can easily be found. However, the problem remains NP-complete for cyclic graphs. To obtain an optimal schedule for cyclic graphs, Hsu suggested that exhaustive searching be performed on all the nodes belonging to nontrivial strongly connected components of the graph [31].

The data storage at each crosspoint has an unusual accessing scheme. It is partly a queue, and partly a register file: data can be read from any location in the storage; however, a write operation always writes to the end of the queue. In this way, addresses need to be specified only for read operations. There are two flavors of read operations; one reads the data and leaves the queue intact, and the other reads the data and deletes it from the queue. When the latter is specified, the queue is automatically compacted by moving all the data after the empty space forward by one location. This design implements modulo variable expansion without having to unroll the object code. Multiple instances of a variable are stored in the queue; they move up the queue as the queue is compacted and all instances of the variable occupy the same location by the time they are used.

The major argument for using the polycyclic crossbar is that optimal results can be obtained for acyclic loops. However, empirical results reported in Chapter 7 indicate that our scheduling heuristics can produce near-optimal schedules without incurring a heavy penalty in compilation time. In particular, most of the acyclic loops can be scheduled optimally in the first attempt in the scheduling algorithm. Moreover, with the polycyclic crossbar, exhaustive search on the nodes of all nontrivial connected components is still necessary to produce the optimal schedule. As we shall show in the next chapter, software pipelining can be extended to loops containing conditional statements. As software pipelining is applied across basic blocks, the size of the problem can become quite large and exhaustive search is unacceptable.

The second advantage of the polycyclic interconnect is that loop un-rolling is not necessary. However, the analysis in Section 5.5 shows that the code size increase is tolerable. Considering the complexity of the interconnect and the width of the instruction word necessary to control the part, unrolling may be more favorable.

While the storage in the crossbar is designed to avoid unrolling, it does not have the full functionality of a register file. It is not possible to write to any random location in the queue, so it is difficult to implement global register allocation such as storing a variable in a register for the duration of an entire loop. For example, we may want to change the value of a variable within a conditional statement. Therefore, we need to write to a specific location, and not just the last location of a queue.

The major drawback of this approach is the cost of the interconnect. This interconnect requires the design of a custom VLSI component. Moreover, the data storage at every crosspoint of the crossbar requires numerous control and addressing signals; these signals imply an ex-tremely wide micro-instruction word and thus a large micro-store. Given that our empirical results show that software heuristics can be used to generate near optimal code, the use of this specialized hardware does not seem to be justified.

5.7. Chapter summary

Software pipelining is an attractive solution to scheduling loops for highly parallel and pipelined data paths because it offers a possibility of achieving optimality with highly compact object codes. There are two intrinsic difficulties in using the technique. First, minimizing the initia-tion interval is an NP-complete problem; the scheduling constraints are defined in terms of the initiation interval, but the feasibility of an initia-tion interval is best shown by a schedule. Second, the flow graph may contain cycles due to inter-iteration data dependencies. The cycles in the graph make it difficult to design efficient non-backtracking heuristics.

These two difficulties have probably discouraged many practitioners from using this technique.

This chapter described the algorithm for scheduling both acyclic and cyclic graphs and discussed the use and implementation of the modulo variable expansion optimization. The results are that near-optimal, and often times optimal, schedules are obtained using the heuristics. They show that software pipelining is a feasible and effective scheduling technique for parallel and pipelined data paths. Good performance can be obtained by software scheduling heuristics without the need for expensive hardware.

6
Hierarchical Reduction

Hierarchical reduction is a unified approach to scheduling both within and across basic blocks. The motivation for the technique is to make software pipelining applicable to all innermost loops, including those containing conditional statements. Software pipelining has previously been defined only for loops containing straight-line code bodies. A simple conditional statement in the loop would have rendered the technique inapplicable. A loop such as the following would have been extremely inefficient on a heavily pipelined and parallel data path.

```
FOR i := 0 TO n DO
BEGIN
    IF a[i] > t THEN
    BEGIN
        count := count + 1;
    END;
END;
```

While the primary performance improvement comes from software pipelining innermost loops, there are many other global code motions that are also desirable. For example, compacting the code within an iteration of an innermost loop is important if it consists of a large number of basic blocks and operations. Or, for innermost loops with few iterations, overlapping the prolog and epilog with scalar operations outside the loop can minimize the penalty of short vectors.

Hierarchical reduction supports all these various global code motions by a uniform mechanism. In this approach, we schedule the program hierarchically, starting with the innermost control constructs. As each construct is scheduled, the entire construct is reduced to a simple node representing all the scheduling constraints of its components with other constructs. This node can then be scheduled just like a simple node within the surrounding control construct. By capturing all scheduling constraints between the nodes, the whole range of local and global code motions is available to the scheduler. The scheduling process is complete when the entire program is reduced to a single node.

Various global scheduling, or microcode compaction techniques, have been proposed in the past. The hierarchical reduction technique is derived from the scheduling scheme previously proposed by Wood [60]. In Wood's approach, scheduled constructs are modeled as black boxes taking unit time. Operations outside the construct can move around it but cannot execute concurrently with it. Here, the resource utilization and precedence constraints of the reduced construct are visible, permitting it to be scheduled in parallel with other operations. This is essential if we want to software pipeline loops with conditional statements effectively.

Dasgupta [12] and Tokoro et al. [55] suggested global compaction be performed as a separate phase after the basic blocks have been individually compacted locally. The locally compacted code is matched against a precompiled menu of code motions, and legal code motions are performed whenever advantageous. As Fisher argues, if each block is compacted separately, too many arbitrary decisions are made during a block's compaction that adversely affect the ability to move operations from block to block [19]. Moreover, each time an operation is moved, it opens up the possibility of more code motion. This technique is more suitable for traditional horizontal microcode engines that have much less parallelism than a high-performance processor like the Warp cell.

Fisher's trace scheduling [19] is another global scheduling technique in which the focus is on constraints of code motion. In trace scheduling, all operations from all basic blocks along the most frequently executed trace are compacted as if they belong to one basic block. All branches are handled the same way; while trace scheduling is tuned for loop branches, it does not handle conditionals well.

A technique commonly used to exploit the potential of data paths with highly pipelined functional units is vector processing. Our approach of directly translating applications into microcode bypasses the intermediate abstractions of vector instructions. Since the microcode generated by the scheduling techniques here are competitive with hand optimized code, this method greatly increases flexibility without losing efficiency.

The organization of this chapter is as follows. We first discuss how iterative and conditional constructs are handled. In the discussion, we assume that the components in the construct are all simple micro-operation sequences. We show how each type of construct, after it has been scheduled, can be represented no differently from a micro-operation sequence. Therefore, scheduling can be applied hierarchically, starting with the innermost constructs. Next, we show the various global code motions that are supported by this unified model. Finally, we conclude the chapter with a comparison with trace scheduling and the vector processing approach.

6.1. The iterative construct

Let us first describe how an iterative construct is represented after it has been software pipelined. The goal of the representation is to allow the construct be scheduled with other operations in the surrounding construct just like a straight-line micro-operation sequence inside a basic block. The representation of the node must include all necessary scheduling constraints to ensure only legal global code motions can be performed. Besides the usual resource and data dependency constraints, loops also have control constraints. The body of a loop in the object code cannot be overlapped with operations outside the loop. That is, no operations can be moved into the steady state of a software pipelined loop. The prolog and epilog, however, can be overlapped with operations outside the construct.

A software pipelined loop is modeled as a straight-line sequence of instructions, whose steps correspond to those taken when the steady state of the loop is executed exactly once. Figure 6-1 gives an intuitive picture of how the resource, data dependency and control constraints are represented in the loop node. In the figure, a dashed rectangle represents a control construct, enclosing all the operations within the construct. Each node is represented by its resource reservation table; precedence constraints are represented by arcs linking the tables.

Resource constraints. The resource requirements of the prolog and epilog can be determined by summing the resource requirements of the different instantiations of iteration in them.

Data dependency constraints. All unsatisfied precedence constraints incident on the nodes within the loop must be transferred to the iterative construct itself. To simplify the procedure, the worst case is assumed: If an operation in any of the iterations must follow some other node, we ensure that the operation in the first instantiation of the loop body obeys this constraint. Similarly, if an operation in any of the itera-

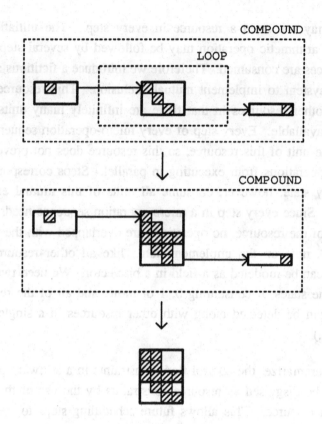

Figure 6-1: Scheduling a loop in parallel with other operations

tions must precede some other, we ensure that the operation in the last instantiation of the loop body obeys this constraint.

Control constraints. The steady state cannot be scheduled to execute in parallel with any other sequences of operations. Mutual exclusion between the iterative portion of the loop and all other operations can be expressed in the form of resource constraints. If every step of every micro-operation sequence requires at least one resource, then if the steady state segment is defined to use every resource in the system, mutual exclusion is guaranteed.

Due to the pipelining of functional units, a micro-operation sequence may not need a resource in every step. The initiation of a pipelined arithmetic operation may be followed by several steps where no resources are consumed. Therefore we introduce a fictitious resource into the system to implement mutual exclusion. This resource differs from all other resources in that there are infinitely many units of this resource available. Every step of every micro-operation sequence consumes one unit of this resource, so this resource does not prevent non-iterative operations from executing in parallel. Steps corresponding to the steady state, however, consume all units of this mutual exclusion resource. Since every step in a micro-operation sequence needs at least one unit of the resource, no operations are overlapped with the iterative code. (A note on the implementation: like all other resources, this resource can be modeled as a field in a bit vector. We need two bits to encode the states of consuming 0, 1 or more, and all of the resources; conflict can be detected along with other resources in a single logical operation.)

To summarize, the control flow constraints in a software pipelined loop can be disguised as resource constraints by the use of the mutual exclusion resource. This allows future scheduling steps to treat loops just like any other straight-line code sequences. General techniques, like list scheduling, can be applied. Suitable code motions, such as scheduling operations outside the loop to use the idle resources in the prolog, are automatically performed as scheduling proceeds.

6.2. The conditional construct

Because conditional statements are difficult to optimize, we concentrate on preventing conditional statements from serializing the execution of other operations. In particular, the presence of conditional statements in a loop must not prevent the loop from being software pipelined. When software pipelining a loop with conditional statements, a conditional statement may execute in parallel with operations in the same or different iterations.

Figure 6-2 illustrates the procedure for handling conditional statements. The THEN and ELSE branches of a conditional statement are first
scheduled independently using list scheduling and padded to the same
length. The entire conditional statement is then reduced to a single node.
This node has, for its scheduling constraints, the union of the scheduling
constraints of the two branches. This node can be treated within the
surrounding construct like any other simple node, since any operation
that can be scheduled with the union of the constraints can be scheduled
with either of the two branches. At code emission time, two sets of code
of equal length, corresponding to the two branches, are generated. Any
code from outside the conditional statement scheduled in parallel is
duplicated in both branches.

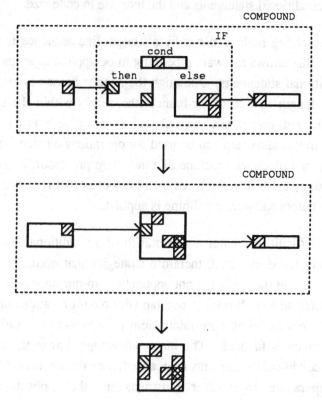

Figure 6-2: Scheduling a conditional statement

The union of the resource requirements for the two branches is computed by finding the maximum of the number of each resource used in each step. Precedence constraints between operations inside the branches and those outside must now be replaced by constraints between the node representing the entire construct and those outside. The attributes of the constraints remain the same.

If the branches of n conditional statements are overlapped together, then 2^n instructions must be generated per step to cover all the possible combinations of the outcomes of the tests in the conditionals. Compacting each branch first allows us to find the shortest schedule possible for each conditional construct; this minimizes the chances of overlapping multiple conditional statements and the increase in code size.

Representing both branches by a straight-line sequence is a simple technique that allows software pipelining to be applied to loops containing conditional statements. Although the shorter branch is extended to the same length as the longer branch, the clocks within the extension period are not completely wasted. Any resource unused by both branches in the same step can be used for operations outside the branch. Typically in a pipelined machine like the Warp processor, it is common to have many more idle resources than busy ones in a resource function of a node before software pipelining is applied.

It is difficult to predict which branch of a conditional statement is more likely to be executed, therefore strategies that optimize a branch while penalizing the other are not applicable. In our approach, the conditional statement is allowed to overlap with other operations only if the union of the scheduling constraints from both branches is satisfied, so neither branch is favored. The major advantage, however, is that the effect of conditional statements is localized; they do not disrupt the overlap of operations from other constructs and the application of the software pipelining technique.

6.2.1. Branches taking different amounts of time

The proposed strategy is designed especially to handle short conditional statements in innermost loops for machines with plenty of parallel hardware. The assumption made is that it is more profitable to satisfy the union of the scheduling constraints of both branches all the time so as not to reduce the opportunity for parallelism among operations outside the conditional statement. The assumption is not true, for example, if the long branch uses almost all the resources in its entire schedule and the other branch is very short. In particular, if we know that the long branch is taken only under exceptional circumstances, then we do not want the schedule to allow for the exceptional cases all the time. Here we discuss how we can extend the technique to allow branches to take different amounts of time.

We can control the inefficiency for having to satisfy the union of the constraints by disallowing operations inside conditional statements to overlap with those outside. Parallelism is reduced within the branches, but code motions around the construct are still allowed. This can be implemented by modifying the schedule of the branches to consume all units of the mutual exclusion resource introduced earlier. The representation of the branches is still the same as before, allowing software pipelining to be applied, if desired. When we emit the code for the branches, instructions containing no micro-operations (and not part of an indivisible sequence) are removed. By removing all resources explicitly, we avoid scheduling small numbers of operations with the branches, so the short branch remains short.

In both the original form and in this extension of the technique, the schedule for the code outside the branches is the same irrespective of the branch taken. This approach greatly simplifies the scheduling procedure.

6.2.2. Code size

The strategy of scheduling the branches of a conditional construct as compactly as possible is aimed at reducing the code size. For every conditional construct, two sets of instructions must be generated: one for each branch. When multiple conditional statements are executed concurrently, the code size increases exponentially with the degree of overlap. In software pipelining, if the length of the branches is greater than the initiation interval, then the same conditional construct from different iterations are overlapped. By compacting the branches first, we try to reduce the lengths of the branches, hence the increase in code size.

In software pipelining, scheduling branches as compactly as possible first is in conflict with the goal to minimize the initiation interval, as discussed in Chapter 5. First, it is harder to pack nodes that use large numbers of resources. Second, by packing the branches first, the schedule of a conditional node may have already violated the resource constraints for some given initiation interval. This can happen if the length of the branches is greater than the initiation interval, and that the modulo resource requirement of the node is greater than the available resources. But, as described above, this is also exactly when the code size increase can become a serious problem. Therefore, it is more important that the branch length is minimized.

Code explosion can be further controlled in other ways as well. First, side-effect free operations, such as the computation of an intermediate expression, may be moved outside the conditional statements to reduce the amount of processing within the branches. Second, we can restrict the number of overlapping conditional branches by another resource. There are n units of this "conditional construct" resource, where n is the maximum number of conditional statements that can be overlapped. After a branch of a conditional statement is scheduled, the resource requirement of the node representing the entire branch is modified by adding one unit of this special resource to the resource usage of each step in the instruction sequence, unless it already consumes all units of the resource. The scheduler automatically enforces the limit in overlapping conditional statements in its process of avoiding resource conflicts.

6.3. Global code motions

The procedures described above allow us to represent entire control constructs as simple nodes after they have been scheduled. Once represented in such a manner, the constructs can be scheduled with other operations in the surrounding control constructs using scheduling techniques previously applicable only to basic blocks. In the W2 compiler, two scheduling algorithms are used: software pipelining is applied to all innermost loops, and list scheduling is used everywhere else. The scheduling process is performed hierarchically, starting with the innermost constructs, and is complete when the entire program is reduced to a single node.

Rearrangement of nodes representing code from different basic blocks corresponds to global code motion. By capturing all scheduling constraints between the nodes, the whole range of local and global code motions is available to the scheduler. Performing code motion at code scheduling time is obviously desirable because all relevant information to derive a good schedule is available at the same time. The following is a list of possible rearrangements of code that can take place using this approach.

1. **Rearrange a conditional statement and operations around it.** There are several possibilities:

    a. *Overlap the branches of a conditional statement with operations outside.*

    b. *Move operations around a conditional construct.*

    c. *Insert operations between the evaluation of the test and the branch operation.* In the Warp architecture, the test in the conditional statement must be initiated many cycles before the result is available for making a branch decision. This is similar to delayed branching [29], where the branch instruction is issued but the control flow is not transferred to the branch until one or more instructions later. The cycles between the evaluation of the test and the branch instruction are treated just like other idle slots created by pipelining in the hardware. Any node that is not restricted by resource or precedence constraints can be scheduled in these slots. They may be from the same basic block, or other basic blocks. Note that operations within the branches of the conditional statement are not eligible because the branches are constrained to execute after the evaluation of the test.

d. *Insert a conditional construct within a long indivisible sequence of micro-operations.* Because of the long pipeline delay, it is possible that entire branches of a conditional statement can be embedded within some indivisible sequence of micro-operations. For example:

```
a := a+1;
if b > c then begin
    b := c;
end;
```

While the assignment of **a** takes 7 cycles on the Warp machine, the split and the rejoin of the second statement can be completed between the initiation of the addition and the assignment.

2. **Overlap multiple conditional statements.** Multiple conditional constructs unconstrained by data dependency may be overlapped just like other micro-operation sequences. An example of a program where two conditional statements may be overlapped is as follows:

```
if a > b then begin
    acount := acount + 1;
end;
if c > d then begin
    c := d;
end;
```

The second conditional statement can be executed "for free" by using the otherwise idle resources when executing the first statement. Emitting code for overlapped conditional statements is slightly more complicated. As suggested in Figure 6-3, the emitted code is not necessarily block-structured. The conditional constructs in Figure 6-3(a) are staggered by one clock unit, and the resulting code is shown in Figure 6-3(b). The arrows in the figure indicate possible control flow.

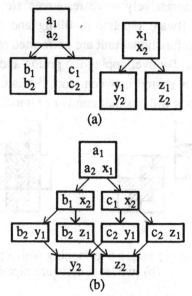

(a)

(b)

Figure 6-3: (a) Staggered conditional constructs, and (b) combined code

3. **Software pipeline a loop with conditional statements.** Since the conditional node is no different from a node representing a straight-line code sequence, software

pipelining can be applied to loops with conditional statements without modification.

4. **Overlap the prolog and epilog of a software pipelined loop with operations outside the loop.** This code motion is especially important if the number of iterations is small. It utilizes the idle resources during the filling and draining of the software pipeline, and minimizes the penalty typi-cally associated with short vectors.

5. **Overlap the epilog with the prolog of the next loop.** This is an important special case of the above. The resource usage pattern of the epilog is generally com-plementary with respect to that of the prolog. As shown in Figure 6-4(a), progressively fewer resources are used in the epilog, but progressively more resources are used in the prolog. In software pipelining, filling and draining the pipelines in the functional unit are performed only once for the entire loop. By overlapping the prolog and epilog, this one time cost is shared between loops. Again, this is sig-nificant for loops with a small number of iterations.

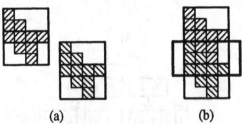

(a)   (b)

Figure 6-4: (a) An epilog overlapping with a prolog, and
(b) two level software pipelining

6. **Software pipeline an outer loop.** Since a software-pipelined loop is represented no differently from any other nodes, software pipelining can be applied to outer loops as well. If the innermost loop has only a small number of iterations, the time spent executing the prolog and epilog may be significant compared to the time spent in the steady

state itself. In that case, it may be advantageous to software pipeline the outer loop as well.

The resource usages of the prolog and epilog are naturally complementary. The epilog of a loop can overlap completely with the prolog of the next instance of the loop; the resource utilization when the prolog and the epilog are overlapped is similar to that of the steady state in the original loop, as depicted in Figure 6-4(b). The steady state of the second software pipelined loop is enclosed by a box of bold lines.

6.4. Comparison with previous work

When comparing different techniques and results in global compaction, it is important to know the underlying machine models. Although the degree of pipelining and parallelism in the hardware does not affect the formulation of the global compaction problem, it determines the efficacy and suitability of the techniques used. We distinguish between two classes of machines: conventional data paths microprogrammed to emulate higher level machine architectures, and data paths consisting of large numbers of functional units for which custom microcode is generated for each application. Early global compaction work [12, 55, 60] described in the introduction of this chapter is targeted for the first category of machines; in these machines, there may be typically three or four units that can be controlled independently. The second kind of machines is epitomized by Yale's VLIW (very long instruction word) architectures [20, 24]. An example configuration of the architecture consists of 8 clusters, each containing two ALUs and two pipelined floating-point arithmetic units [15]. The Warp cell with its two 7-staged pipelined arithmetic operations and various memory storages also belongs to this class.

The fundamental difference between the microprogrammed macro architectures and VLIW class machines is that data dependency for the

latter can result in extremely sparse schedules if basic blocks are compacted locally. While rearranging operations from neighboring basic
blocks is a good solution to the global compaction problem for the first
kind of machines, the technique is not adequate for the second.
Similarly, techniques allowing more global code motions may be un-
necessarily expensive for the first. Results on experiments of global
compaction must be interpreted in light of the underlying machine
model.

In this section, we concentrate on the similarities and differences of
our approach with other techniques intended to be used for processors
with multiple parallel and/or pipelined function units. We first compare
our approach to trace scheduling, then to vector processing. The vector
processing approach is interesting because the characteristics of the functional units in each Warp cell are similar to those of vector machines.
We will contrast our approach of custom generation of microcode from a
high-level language with the use of a vector instruction set.

6.4.1. Trace scheduling

The primary idea in trace scheduling [19] is to optimize the more
frequently executed traces. The procedure is as follows: first, identify
the most likely execution trace, then compact the instructions in the trace
as if they belong to one big basic block. The large block size means that
there is plenty of opportunity to find independent activities that can be
executed in parallel. The second step is to add compensation code at the
entrances and exits of the trace to restore the semantics of the original
program for other traces. This process is then repeated until all traces
whose probabilities of execution are above some threshold are scheduled.
A straightforward scheduling technique is used for the rest of the traces.

The strength of trace scheduling is that all operations for the entire
trace are scheduled together, and all legal code motions are permitted. In

fact, all other forms of optimization, such as common subexpression elimination across all operations in the trace can be performed. On the other hand, major execution traces must exist for this scheduling technique to succeed. In trace scheduling, the more frequently executed traces are scheduled first. The code motions performed optimize the more frequently executed traces, at the expense of the less frequently executed ones. This may be a problem in data dependent conditional statements. Also, one major criticism of trace scheduling is the possibility of exponential code explosion [32, 39, 42, 51].

The major difference between our approach and trace scheduling is that we retain the control structure of the computation. By retaining information on the control structure of the program, we can exploit the semantics of the different control constructs, control the code motion and hence the code explosion. Another difference is that our scheduling algorithm is designed for block-structured constructs, whereas trace scheduling does not have similar restrictions. The following compares the two techniques in scheduling loop branches and conditional branches separately.

6.4.1.1. Loop branches

Trace scheduling is applied only to the body of a loop, that is, a major trace does not extend beyond the loop body boundary. To get enough parallelism in the trace, trace scheduling relies primarily on source code unrolling. At the end of each iteration in the original source is an exit out of the loop; the major trace is constructed by assuming that the exits off the loop are not taken. If the number of iterations is known at compile-time, then all but one exit off the loop are removed.

Software pipelining is more attractive than source code unrolling for two reasons. First, software pipelining offers the possibility of achieving optimal throughput. In loop unrolling, filling and draining the hardware pipelines at the beginning and the end of each iteration make optimal performance impossible. The second reason, a practical concern, is perhaps more important. In trace scheduling, the performance almost always improves as more iterations are unrolled. Finding the suitable degree of unrolling for a particular application often requires experimentation. As the degree of unrolling increases, so do the problem size and the final code size.

In software pipelining, the *object* code is sometimes unrolled to implement modulo variable expansion, but the unrolling is performed at code emission time. Therefore, code scheduling time is unaffected. Furthermore, unlike source unrolling, there is an optimal degree of unrolling for each schedule, and can easily be determined after scheduling.

Software pipelining does not preclude source code loop unrolling. It is sometimes useful to unroll loops for reasons other than exposing parallelism to the compiler. A traditional reason to unroll loop is to reduce the overhead of loop control. If the machine instruction set does not support parallel branching, then we can reduce the cost of loop control by unrolling the loop a small number of times. (The unrolled loop body must not contain any exits out of the loop. Otherwise, a branch per iteration in the original loop is still necessary.) Loop unrolling is sometimes necessary for processors that have multiple units of identical resources. For example, if the loop body contains only a single addition and the machine has several adders, the machine can potentially execute the loop at several iterations per clock. Software pipelining can deliver at most one iteration per clock. This problem can be solved by unrolling the loop several times.

The idea of "software pipelining" has been considered in the context of trace scheduling. Fisher suggested unrolling loops as needed until a repeating pattern was observed in the object code [23]; the repeating pattern corresponds to the steady state in software pipelining. This technique is not practical because the "steady state" obtained by this approach is likely to be very large. A practical adaptation of this idea is suggested and implemented by Su et al. [51]. Their procedure is to unroll the loop once, trace schedule the two iterations, and then reroll the loop into one. Rerolling is achieved by examining the code to locate the smallest sequence of steps that cover all the operations in one iteration. After removing redundant operations from the schedule, we obtain a steady state of the loop. This technique is a much simplified version of software pipelining, suitable only for machines with much lower degrees of concurrency. Moreover, it is unclear how this technique can be extended to handle conditional branching within a loop.

6.4.1.2. Conditionals

In the case of data dependent conditional statements, the premise that there is a most frequently executed trace is questionable. While it is easy to predict the outcome of a conditional branch at the end of an iteration in a loop, outcomes for all other branches are difficult to predict. In a study on frequencies of branches [49], branches are classified according to whether the branch target address is above or below the current instruction; almost all the branches belonging to the former class are loop branches. Results showed that for the latter class of branches, taking or not taking a branch is equally likely.

The generality of trace scheduling makes code explosion difficult to control. Some global code motions require operations scheduled in the main trace to be duplicated in the less frequently executed traces. Since basic block boundaries are not visible when compacting a trace, code motions that require large amounts of copying, and may not even be significant in reducing the execution time, may be introduced.

Ellis shows that exponential code explosion can occur by reordering conditional statements that are data independent of each other [15]. Massive loop unrolling has a tendency to increase the number of possibly data independent conditional statements. Code explosion can be controlled by inserting additional constraints between branching operations.

For example, Su et al. suggested restricting the motions of operations that are not on the critical path of the trace [51].

In our approach to conditional statements, the objective is to minimize the effect of conditional statements on parallel execution of other constructs. By modeling the conditional statement as one unit, we can software pipeline all innermost loops. The resources taken to execute the conditional statement may be as much as the sum of those of both branches. None of the operations outside the conditional statements are affected. The increase in code size due to overlapped conditional constructs can be controlled, as described in Section 6.2.2.

6.4.2. Comparison with vector instructions

Vector instructions have been a useful abstraction for managing data paths with highly pipelined functional units. The user, or a compiler, expresses the algorithm in terms of vector instructions, and the hardware manages the low level parallelism in the heavily pipelined data path. Like a vector processor, the Warp cell also has multiple highly pipelined floating-point functional units. However, it does not have a vector instruction set; microcode is directly generated from the user's application program written in a high-level language.

The alternative of microcode compilation is now made feasible by sophisticated scheduling techniques such as software pipelining. As shown in the next chapter, software pipelining can deliver performance that is comparable to hand-crafted microcode. One advantage that vector instructions have over microcode compilation is the small code size. Provided that the machine has sufficient program storage, microcode compilation offers various advantages, as described below.

In general, vector instructions increase the need for buffers to hold intermediate vectors of data. For example, a simple innermost loop of

```
FOR i := 0 TO n DO BEGIN
    d[i] := a[i]+b[i]+c[i];
END;
```

requires the computation be broken down into two vector-add instructions. If there is only one adder in the data path, chaining cannot be applied and the vector of partial sums **a[i]+b[i]** must be buffered.

Custom generation of microcode is much more flexible and adaptive to the user's program. It alleviates the need for intermediate buffering of vectors of data. For example, the microcode generated for the innermost loop above will produce a result every two cycles; the one adder in the machine is time multiplexed to implement two "vector-add" instructions. There is no need to buffer an entire vector of partial sums; as each partial sum is generated, it is routed back to the adder for the second add operation. Moreover, the latency of the operation is reduced. The first result is ready and sent to the next cell even before the last set of data arrives at the cell.

Moreover, efficient code can be generated for programs that are hard to vectorize. Software pipelining can support general recurrences in a loop, as well as conditional statements. Lastly, scalar and iterative constructs can be easily intermixed, thus significantly reducing the penalty associated with the start up of short vectors.

7
Evaluation

The ideas and techniques in this book have been validated and implemented in the W2 compilers for the Warp machines. The original W2 compiler built for the first prototype of the Warp machine was modified and retargeted as the Warp architecture evolved. The compiler was first released in late 1985 for the prototype Warp machine. Large numbers of applications including robot navigation, low-level vision, signal processing and scientific computing have been developed using this compiler by early 1987 [3, 4]. The compiler has since been retargeted to the PC production machine and iWarp, the integrated Warp architecture. While the PC machine is similar to the first prototypes, the iWarp machine is significantly different. The different compilers contain different machine-dependent optimizations but they all use the same scheduler module.

This chapter presents two sets of evaluation data: a careful analysis of performance of a large set of user programs on the first prototype Warp machine, and performance numbers for the Livermore Loops on the PC Warp. The former analysis was performed in early 1987 before the arrival of the PC Warp. The PC machine is upward compatible with the wire-wrapped prototype. Application programs developed on the prototype run at about the same speed on the PC Warp, so we did not repeat the same analysis on the PC Warp. However, the PC Warp machine has a much greater application domain. Since the arrival of the PC Warp. many new applications have been developed and evaluated. For example, the NETtalk neural network benchmark [48] runs at 16.5 million connections per second or 70 MFLOPS on a 10 cell array [45]. Some Livermore Loop benchmark numbers on the PC Warp are presented in this chapter.

7.1. The experiment

In the analysis of user programs, a sample of 72 user programs was collected and analyzed. The overall average of the computation rate for these programs is 28 MFLOPS, that is, 28% of the peak rate of the array. We have performed experiments to study the factors governing the performance of the programs and the efficiency of the scheduling techniques. We studied the performance of entire programs, as well as the innermost loops, since they are crucial to the efficiency of a program.

Here we first present performance measurements on entire programs to provide a picture of the overall effect of the global scheduling techniques. We compare the performance of the code obtained with both a lower and an upper bound. Code generated without using any global scheduling techniques is used to provide the lower bound; comparison with this bound yields the measurement of improvement achieved by the scheduling techniques. The results are that an average speed up of three times is observed, incurring an average of 30% increase in code size. A theoretical upper bound on the performance is calculated from resource and data dependency considerations; comparing the execution times of the programs with this bound gives us the efficiency of the scheduling techniques. The results are that, on average, the efficiency is within 20% of the optimal.

We next zoom in and study the performance of software pipelining on innermost loops. We study the characteristics of the loops, measure the effectiveness of the software pipelining heuristics, and also examine various other parameters in the generated code, such as the usage of registers and the degree of unrolling. We again compare the performance of the innermost loops with a lower and an upper bound. For a lower bound, we compile the loops using only hierarchical reduction and not software pipelining; empirical results show that software pipelining improves the code by a factor of four on average, and that pipelining loops with conditional statements is feasible and is important. Comparing the performance with the upper bound obtained from resource and data dependency considerations shows that optimal throughput is achieved for at least 73% of the loops.

Before analyzing the performance data, let us first present some background material relevant to the interpretation of the data: the implementation of the compiler and the description of the programs.

7.1.1. Status of compiler

The focus in the development of the W2 compiler has been on code scheduling techniques. All the algorithms described in this book have been implemented. The W2 compiler lacks those high-level optimizations that are commonly used in vectorizing compilers. These optimizations can be applied by the user at the source level, and most programs in the sample have indeed been optimized in this manner. As these source-to-source transformations are outside the domain of the scheduler, we assume that these optimizations have already been applied when calculating the lower bound on the execution time of the program on the machine.

Loop fusion [33] is one of the optimizations commonly used in vectorizing compilers that is also applicable to this compiler. Fusing two or more loops into one loop increases the number of independent operations scheduled together, and may lead to a faster program. For example, consider the case when one loop uses only the I/O resources and another uses only the arithmetic units. Since the I/O operations can be executed concurrently with the arithmetic operations, the execution time of the code generated for the combined loop may equal to the maximum, rather than the summation, of the execution times of the individual loops.

Another example of a useful high-level transformation is loop interchanging [58]. Since the efficiency of the innermost loops dictates the efficiency of the program, it is important that the computation of the innermost loops be able to make use of the parallel hardware. For example, the following loop

```
FOR i := 0 TO n DO BEGIN
    FOR j := 0 to n DO BEGIN
        A[i] := A[i] + B[i,j];
    END;
END;
```

sums up the ith row of B in A[i]. Since each iteration of the innermost loop needs the result of the previous iteration, different iterations cannot be overlapped. On the other hand, if the loops are interchanged as follows:

```
FOR j := 0 to n DO BEGIN
    FOR i := 0 TO n DO BEGIN
        A[i] := A[i] + B[i,j];
    END;
END;
```

The same computation is performed, but the computations in different iterations of the innermost loop are independent and can thus be overlapped.

7.1.2. The programs

The programs used in the evaluation are a sample of actual applications developed on the Warp prototype machine. Because of the restrictions on the prototype machine, these programs are subjected to the restrictions imposed by compile-time flow control. That is, they all have compile-time constant loop bounds and the conditional statements do not contain loops. The code generated for the two branches of conditional statements must execute in same amounts of time. These restrictions are no longer present in the PC Warp machine, and the compiler handles all arbitrary nesting of loops and conditional statements. All innermost loops, including WHILE loops and FOR loops of dynamic loop bounds, are software pipelined.

The programs in the experiment are applications in robot navigation, low-level vision, signal processing, and scientific computing [3, 4]. In robot navigation applications, Warp is used to process the massive amounts of low-level input data and generate high-level information for the decision making process running on the general-purpose host. The applications implemented include road following, obstacle avoidance using stereo vision, obstacle avoidance using a laser range-finder, and path planning using dynamic programming. All the W2 programs implemented for these applications are included in the analysis.

The second group of application programs comprises of routines from the SPIDER library [50], a library of functions used in vision research. There are over 100 programs from the SPIDER library implemented to date. Only programs with significantly different performance characteristics are included in the experiment.

Routines in signal processing and scientific computing included in the experiment are: singular value decomposition (SVD) for adaptive beamforming, two-dimensional image correlation using fast Fourier transform (FFT), successive over-relaxation (SOR) for the solution of elliptic partial differential equations, routines for magnetic resonance image processing, as well as computational geometry algorithms such as convex hull and algorithms for finding the shortest paths in a graph.

A one-line description of the sample programs, their execution times and their achieved computation rates, in terms of number of floating-point operations per second, are presented in Table 7-1. Because of the restrictions imposed by the prototype machine as described above, it is possible to calculate the execution times of the programs at compile time. The computation rates are calculated by dividing the total number of floating-point operations in the program by the execution time; branches of a conditional statement are assumed to execute equal number of times.

Performance of specific algorithms on Warp		
Task All images are 512×512.	Time (ms)	MFLOPS
Mandelbrot image, 256 iterations.	6968	88.1
100×100 matrix multiplication.	25	79.4
512×512 complex FFT (per dimension).	164	71.9
3×3 convolution.	70	65.7
11×11 symmetric convolution.	367	58.8
100×100 singular value decomposition (one pass).	153	58.2
Calculate transformation table for non-linear warping.	248	57.1
Iterative enhancement of noisy image.	691	49.9
Generate matrices for plane fit for obstacle avoidance using ERIM scanner.	174	49.3
Edge detection using orthogonal templates by Frei and Chen.	301	48.4
One-iteration process for iterative enhancement of noisy images.	1415	45.5

Performance of specific algorithms on Warp		
Task All images are 512×512.	Time (ms)	MFLOPS
7×7 Average grayvalues in square neighborhood with a certain angle.	1766	44.3
Generate mapping table for affine image warping.	225	42.7
Generate diamond pattern.	277	42.4
Hough transform.	2113	42.2
SOR solving linear system of 50625 unknowns (10 iterations).	202	41.0
Edge detection using the Kirsch operator (magnitude and direction).	329	40.3
Edge preserving smoothing.	1215	39.6
Compute connectivity number.	272	39.2
Local selective averaging.	406	39.2
Generate grid pattern.	191	37.6
Generate stripe pattern.	198	36.4
Moravec's interest operator.	82	35.5
Generate bull's-eye pattern.	183	30.6
3×3 maximum filtering.	280	30.3
Sobel edge detection.	206	29.6
Convert expression modes of complex data.	697	27.8
Label color image using quadratic form for road following.	308	27.2
Iterative edge detection using Kasvand's method.	508	27.1
Obtain 0-th to second moments of two-dimensional image.	174	26.2

Performance of specific algorithms on Warp		
Task All images are 512×512.	Time (ms)	MFLOPS
Image magnification using cubic spline interpolation.	8438	25.1
Edge detection using Prewitt operator (differential type) magnitude.	206	24.5
Shrinking a binary image.	135	24.4
Growing a binary image.	135	24.4
Shortest path in 350 node graph using Warshall's algorithm (10 iterations).	104	24.3
Smooth an image while preserving texture edges.	1093	24.1
31×31 Edge value of texture edge of a specified size and direction.	395	23.5
5×5 convolution.	284	23.3
Calculate quadratic form from labeled color image.	134	22.4
Generate random numbers.	375	22.1
Measure coordinates of circumscribing rectangle.	322	21.7
Histogram over a region of an image.	185	21.5
Compute gradient using 9×9 Canny operator.	473	20.8
Discrete cosine transform on 8×8 windows.	177	20.7
3×3 Laplacian edge detection.	228	19.8
Extract or delete points in an image.	177	19.5
Compute crossing number.	225	18.9
Assign constant to inside of irregularly-shaped region.	236	18.3

Performance of specific algorithms on Warp		
Task All images are 512×512.	Time (ms)	MFLOPS
Point symmetry using analysis of variance statistical test.	47655	17.7
Inverse discrete cosine transform on 8× windows.	174	17.7
Find zero-crossings.	179	16.4
Detect borders in binary picture.	199	16.0
Roberts operator.	192	15.2
Solve matrices for plane fit in ostacle avoidance using ERIM scanner.	107	14.3
Compute the logarithm of each pixel in a real image.	468	14.0
Find next convex hull point of 1000 points.	660	13.5
Calculate (x,y,z) coordinates from ERIM laser range scanner data.	24	12.8
Generate checkboard pattern.	180	12.6
Histogram.	67	12.0
Coarse-to-fine correlation for stereo vision.	12	11.1
Scale image and set pixels outside range to constants.	147	10.9
7×7 edge value of texture edge horizontally or vertically.	143	9.3
Requantize image by reducing graylevels.	154	8.6
Measure coordinates of center of gravity.	193	8.5
Apply grayvalue translation table.	66	8.0
3×3 median filter.	448	7.8
Levialdi's binary shrink operation.	180	7.4

Performance of specific algorithms on Warp		
Task All images are 512×512.	Time (ms)	MFLOPS
31×31 average grayvalues in square neighborhood.	444	5.0
Convert real image to integer using max, min linear scaling.	249	4.4
Path planning using dynamic programming (10 iterations).	755	3.7
Image reduction by non-overlapping 2×2 windows.	154	3.0
Best-edge size, direction, and value using average grayvalues.	735	1.1

Table 7-1: Time and arithmetic computation rates of the 72 programs

The distribution of the computation rates is graphically presented in Figure 7-1. The computation rates of the programs are useful for studying the efficiency of the compiler. However, for benchmarking and comparison with other architectures, the execution times of the individual applications are more meaningful. The reason is that the number of operations executed depends on the algorithm, the compiler, and the characteristics of the machine. An algorithm efficient for an array of cells may have more arithmetic operations than that for a sequential processor. An optimizing compiler may reduce the number of operations by, for example, common subexpression elimination.

There are a few programs in the sample for which the computation rates are inflated because of the characteristics of the Warp machine. Computation within a conditional is sometimes hoisted and executed unconditionally to reduce the code size and possibly the execution time of the program, while increasing the count of floating-point operations. Because the cells on the prototype machine do not have an integer ALU, operations normally performed in fixed-point arithmetic are computed on the floating-point units and are also counted as floating-point operations.

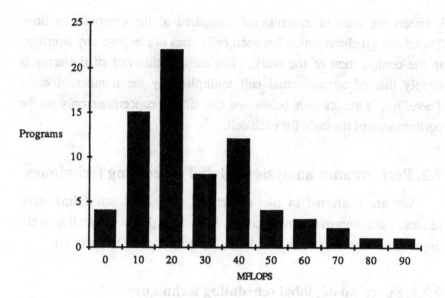

Figure 7-1: Distribution of achieved computation rates

The restrictions that the loop bounds must be compile-time constants and that conditional statements cannot contain loops also increase the arithmetic operation count. For example, efficient calculation of the Mandelbrot set requires the use of a WHILE statement. Although the MFLOPS rate of the program is high, the execution time is not as impressive because it reflects the time it takes to execute the maximum number of iterations for all pixels. We have rewritten the program for the PCWarp using WHILE constructs. The achieved computation rate is 45 MFLOPS, taking $0.2\mu s$ per iteration per pixel. Although the MFLOPS rate is lower, the program is actually much faster in all interesting cases.

All of the seventy-two programs are written for ten cells, with the only exception of the nine-cell program for 512×512 complex fast Fourier transformations. They are all homogeneous programs (all cells execute a copy of the same program), and the data flow across the array is unidirectional. The cells operate in a skewed fashion; that is, each cell starts computing shortly after its preceding cell has started. The skew between the cells is insignificant compared to the computation time. Therefore, synchronization between cells does not impose any overhead in the computation of the array. The computation rate of the array is simply that of an individual cell multiplied by the number of cells. Therefore, in the analysis below, we can simply concentrate only on the performance of the code for each cell.

7.2. Performance analysis of global scheduling techniques

We are interested in two aspects of the global scheduling techniques: the improvement obtained from the global scheduling techniques, and how far are the scheduling techniques from the optimal.

7.2.1. Speed up of global scheduling techniques

To study the effects of software pipelining and hierarchical reduction, the scheduler is modified such that the optimizations can be applied independently. When software pipelining is used without hierarchical reduction, only innermost loops containing straight-line loop bodies are pipelined, and no other code motion is performed. When hierarchical reduction is used without software pipelining, no code motion between iterations is performed. When both global optimizations are disabled, individual basic blocks are simply locally compacted, and no code motion across any basic blocks is performed. Code produced using only local compaction is used as a baseline for comparison with globally optimized programs.

In this experiment, we compiled all the 72 programs with each combination of global scheduling techniques. Comparing the code generated using all optimizations with locally compacted code gives us a measurement of the overall improvement. The performance of the programs generated with either of the optimizations allows us to study the contribution of the individual techniques and their interaction.

Figure 7-2 shows the distribution of the factors of speed up for the various combinations of optimization techniques. The histograms plot the percentages of programs for each interval of factor of speed up. For each interval, we also show the composition of the type of programs: those that contain conditional statements and those that do not. Of the 72 programs, 42 have conditional statements in them. This classification is interesting because the significance of the scheduling techniques is different for the two groups of programs.

The global optimizations speed up the programs by a factor of 3 on average. We observe that the speed up is more significant for programs containing conditional statements. The reason is that conditional statements break up the program into even smaller basic blocks, making global scheduling techniques more important.

From the graphs in Figure 7-2, we observe that hierarchical reduction has little effect on programs that do not contain any conditional statements. Innermost loops dominate the execution time of these programs. Since the innermost loops of such programs are just single basic blocks, hierarchical reduction has no impact on the innermost loops, and therefore little effect on the overall execution time. Software pipelining is primarily responsible for the improvement in execution time.

For programs whose innermost loops contain conditional statements, hierarchical reduction can speed them up by compacting the innermost loop bodies. If hierarchical reduction is applied without software pipelining, the available parallelism is limited to that within an iteration of a loop. As shown in the figure, the improvement is noticeable but not as substantial as when software pipelining is also applied. If software pipelining alone is used, loops with conditional statements cannot be pipelined and therefore are not improved. Some speed up is still observed for programs with conditional statements because they may contain other loops that do not have conditional statements and can thus be

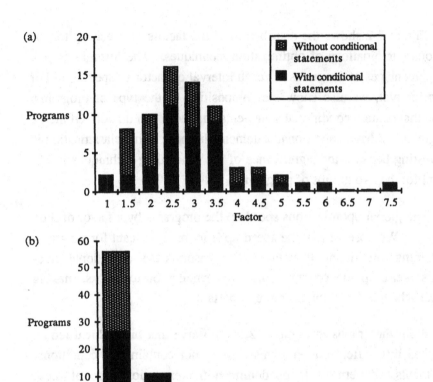

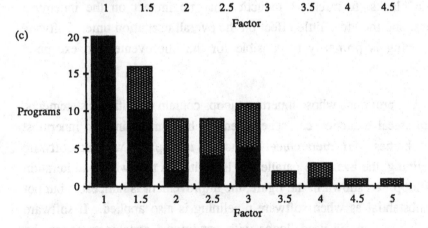

Figure 7-2: Speed up: (a) with all optimizations,
(b) with hierarchical reduction only,
and (c) with software pipelining only

pipelined. If both hierarchical reduction and software pipelining are applied, all innermost loops are pipelined, and a consistent improvement in performance is observed.

Figure 7-3 shows the effect of global scheduling on code size. When compared to code produced using only local compaction, an average increase by 35% is observed. Although compaction of code outside the innermost loop by hierarchical reduction typically has little effect on the execution time, it decreases the object code size by roughly 25%. Software pipelining, on the other hand, tends to increase the program size. The increase is more pronounced when software pipelining is applied to loops with conditional statements; this code expansion happens when multiple conditional statements from different iterations are overlapped.

7.2.2. Efficiency of scheduler

The peak computation rate of the Warp array is 100 MFLOPS; but, on average, the programs studied execute at 28 MFLOPS. How much of this difference between the peak rate and the achieved rate can be attributed to the inefficiency of the scheduler? We have established that global scheduling techniques speed up the programs by an average factor of 3. Here we discuss how close the achieved performance is to optimal.

The efficiency of the scheduler is measured by accounting for the execution time of the programs. There are three major components in this execution time: the exclusive I/O time, the most heavily used resource, and delay caused by data dependencies. Explanation for these components is given below. The graph in Figure 7-4 shows the breakdown of the execution times of the 72 programs.

The execution times of the programs in the figure are normalized, and the programs are sorted in increasing percentage of time accounted for by these three factors. On average, 80% of the computation time is

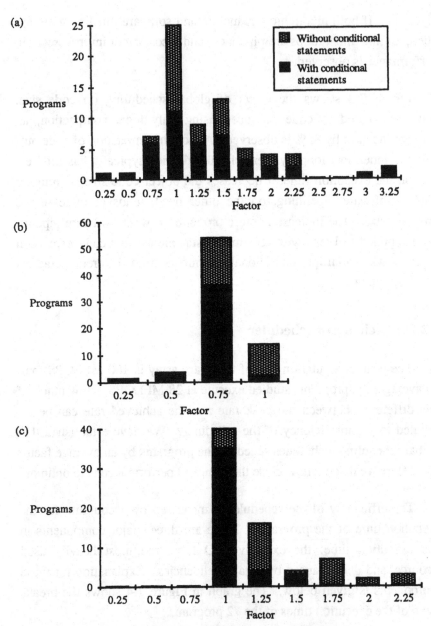

Figure 7-3: Code size increase: (a) with all optimizations,
(b) with hierarchical reduction only,
and (c) with software pipelining only

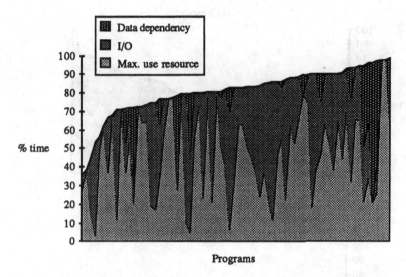

Figure 7-4: Accounting for the execution time of the programs

accounted for, therefore the scheduler's inefficiency is bounded to within the 20% of the total execution time. Figure 7-5 shows the different factors separately. The programs are again sorted in increasing time spent in each factor, and the total contribution of each factor in the execution time of all programs in the sample is given by the area below the curve.

7.2.2.1. Exclusive I/O time

One common way of using the ten cells of the Warp machine is to partition the data and computation across the array, as discussed in Chapter 3. A straightforward algorithm implementing this scheme consists of a loop with three steps: input and distribute data across cells, have each cell compute with its local data, collect and output all data to the host. In such an implementation, the input and output phases are not overlapped with the computation step. The arithmetic units are idle in the I/O phases, and vice versa. The total execution time is equal to the sum of the I/O and the computation phases.

A program with m inputs and m outputs requires $2m$ clock cycles for I/O, regardless of the number of cells in the array. The computation,

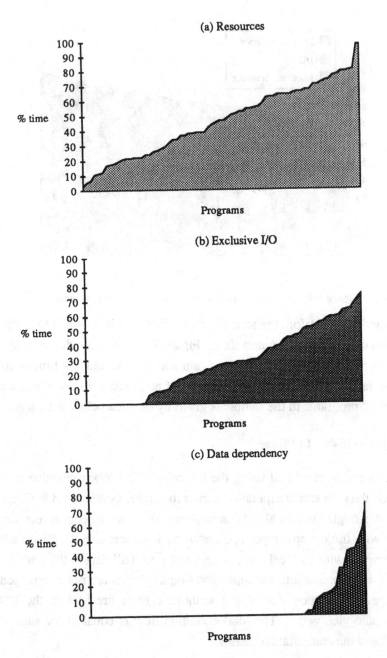

Figure 7-5: Contributions of different factors to execution time

however, is divided across the n cells and takes km/n clock cycles, where k is the number of operations necessary to compute each result. In other words, while the computation time decreases with the number of cells in the system, the I/O time remains constant.

Of the 72 programs, 58 programs are written in the style described above. Many of these are vision programs operating on a fixed size (512×512) image. While a constant amount of time is spent on the I/O; the ratio of I/O to the total execution time decreases as the amount of processing increases. Figure 7-5 (b) shows that the percentage of time spent on I/O can range up to 75%. In the more conventional systolic algorithms, I/O is overlapped with the computation. Fourteen of the programs in the sample are written in this style. They do not have an *exclusive* I/O component and their execution time is the maximum of the I/O and computation time. Overlapping the I/O and computation is especially recommended when the number of operations performed per data item is small.

7.2.2.2. Global resource use count

Accounting for an average of 47% of the execution time is the use of resources. The maximum of the use counts of individual resources in the program gives a lower bound on the execution time. The use count here refers to the normalized count, obtained by dividing the true use count of the resource in the program by the number of units available in each instruction cycle. For example, unless a program has equal numbers of multiplications and additions, either the adder or the multiplier on the cell's data path must lie idle sometimes and 100 MFLOPS cannot be attained. Figure 7-5(a) shows the utilization of the most heavily used resource in each program; the use of resources in the exclusive I/O loops is not included here.

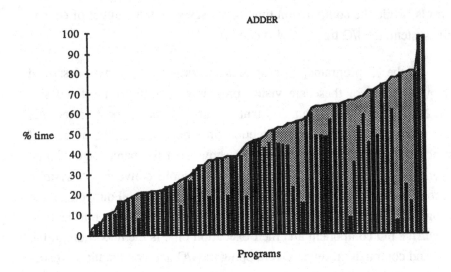

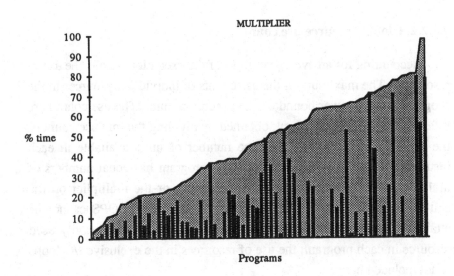

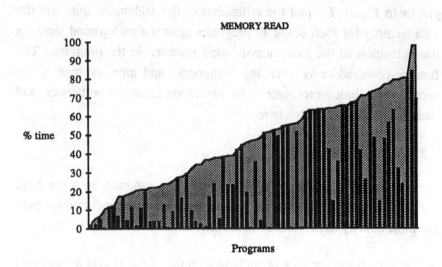

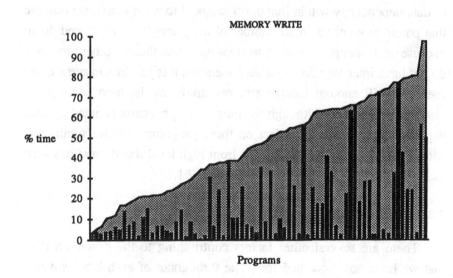

Figure 7-6: Utilization of resources in a Warp cell

There are six major resources in the prototype machine: adder, multiplier, memory reads, memory writes, X queue, and Y queue. The graphs in Figure 7-6 plot the utilization of the arithmetic units and the data memory for each of the 72 programs against a background showing the utilization of the most heavily used resource in the program. The floating-point adder is often the bottleneck, and memory read is the second most limiting resource. The queues are rarely the bottleneck, and their usages are not shown here.

7.2.2.3. Data dependency

On a pipelined and parallel machine, if a program does not have enough independent operations that can be executed concurrently, even the most heavily used resource count may not be utilized all the time.

Figure 7-5(c) shows a lower bound on the effect of data dependency on the performance of the sample programs. It plots the minimum amount of time the resource of maximum use count must be idle because of data dependency within innermost loops. From the graph, we observe that potentially only a small number of programs (16) are slowed down because of data dependency. This does not mean that the other programs do not have inter-iteration data dependencies, it is just that in some cases there are still enough independent operations in the loop to keep the resources occupied. Although the number of programs affected by data dependency is small, the effect on these programs is quite significant. Most of these programs can benefit from high level transformations such as loop interchanging, described in Section 7.1.1.

7.2.2.4. Other factors

There are several other factors contributing to the execution time that we have not accounted for. The throughput of each loop is determined by the resource that is used the most within the loop. The resource utilization shown in the Figure 7-5(a) is a global count, and thus gives only a lower bound on the resource utilization component in the total execution time. Another factor unaccounted for is the effect of conditional control flow. The presence of conditional statements makes it impossible for any compiler to produce a finite sequence of code that is optimized for all data inputs. Therefore, the 20% of the execution time that we have not accounted for is an upper bound on the time lost due to the inefficiency of the scheduler.

7.2.3. Discussion on effectiveness of the Warp architecture

Several observations on the architecture can be made from this analysis. First, although the hardware can sustain a high peak computation rate, the effective rate is governed by the distribution of the resource use counts. For example, there are often many more additions than multiplications in a program. The Warp cell has one adder and one multiplier. Just accounting for the different numbers of additions and multiplications in the sample programs, the performance of the machine can at most be 70 MFLOPS, on average. Sometimes the memory or the I/O units are more heavily used. If all resources are considered, the achievable performance is dropped to an average of 50 MFLOPS. In other words, this value is an upper bound on the performance delivered by any compiler for the sample of programs on the machine, provided that the number of operations executed does not change. Even if there is hardware support to overlap the I/O phase with the computation of the machine, the best attainable performance is still limited to half the peak computation rate, for the sample of programs studied. This observation is, however, *not* an argument against horizontal architectures. Though the peak performance rate cannot be achieved, the obtained performance is still superior to sequential access of each resource.

Second, pipelining is an effective hardware optimization technique for implementing processors of a systolic array. Typically, applications that are suited to implementation on systolic arrays are inherently parallel. The compiler techniques developed in this work allow us to find the parallelism in the code not only within basic blocks, but across basic blocks and iterations in a loop. Removing pipelining from the machine architecture does not have a dramatic effect on most of the programs in the sample; it will, however, reduce the code size of the programs.

7.3. Performance of software pipelining

After describing the overall effects of software pipelining and hierarchical reduction on entire programs, we now further analyze the performance and the feasibility of the software pipelining algorithm by studying the individual loops in the programs. In the following, we first describe the characteristics of the loop, then the experimental results.

7.3.1. Characteristics of the loops

Only innermost loops that use the arithmetic units are included in this study. We filter out all other loops because many of the programs have identical loops where only I/O is performed. These loops are typically very simple and are invariably sped up by a factor of two or three when software pipelined. Altogether, there are 122 loops with arithmetic computation. All but five of them are pipelined successfully. In the following, we first study those that can be pipelined; we will describe the unpipelined ones at the end.

In the following performance evaluation, we distinguish among the loops according to whether they contain conditional statements, and whether they have non-trivial connected components in the flow graph. In the software pipelining algorithm, conditional constructs and non-trivial connected component are individually scheduled first before they are software pipelined with other components in the flow graph. The effect of the scheduling algorithms on these different classes of loops is different.

The breakdown of the composition of the pipelined loops in the experiment is:

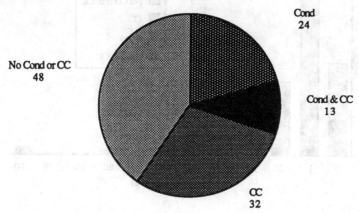

There is a large variety of loops in the sample. Two representative characteristics of the loops are chosen and presented in Figure 7-7 to give a general picture of the sample: the initiation interval, and the time to execute one iteration of the source loop. The former determines the rate at which the iterations are executed. The latter quantity determines the latency in computing one iteration; it is typically only slightly longer than the execution time of one iteration if the loop is not pipelined.

7.3.2. Effectiveness of software pipelining

The throughput at which the iterations in the loop are executed is inversely proportional to the initiation interval. We can measure the effectiveness and efficiency of software pipelining by comparing the initiation intervals obtained for the loops in the sample with a lower and upper bound. For a lower bound on the initiation interval, we use the theoretical minimum initiation interval determined from the resource and precedence constraints, as described in Chapter 5. For an upper bound, we use the execution time of an iteration obtained by applying hierarchical reduction to the loop body.

(a)

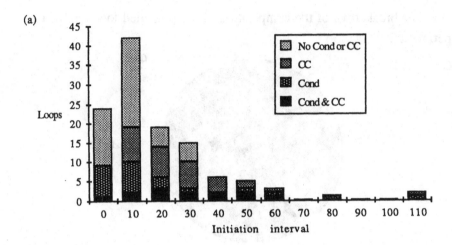

(b)

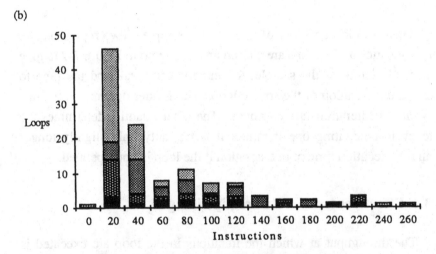

Figure 7-7: Distribution of (a) initiation intervals, and
(b) execution times of single iterations

The theoretical lower bound on the initiation interval is calculated
from the resource requirements and the cycle lengths in the precedence
constraint graphs. In acyclic graphs, the minimum initiation interval is
simply the use count of the most heavily used resource; so if a schedule
meeting the minimum initiation interval can be found, then the most
heavily used resource is busy all the time. In cyclic graphs, this minimum interval value may further be raised by the length of the longest

cycle in the precedence constraints graph. Cycles in the graph do not necessarily imply that there must be non-trivial connected components in the loop, because a one-node cycle would still result in a trivial connected component. On the other hand, all loops with non-trivial connected components must have cycles. About half of the loops containing non-trivial connected components have enough independent operations to theoretically keep the resources busy at all time.

For each loop, we obtain three measurements: (a) the initiation interval of the loop, (b) the execution time of a hierarchically reduced code for one iteration of the loop, and (c) the theoretical lower bound calculated as above. The potential speed up of a loop is the ratio between the execution time of an unpipelined iteration of the loop and the theoretical lower bound; the actual speed up obtained is the ratio between the execution time of an unpipelined iteration and the achieved initiation interval. The distributions of the potential and actual speed ups are shown in Figure 7-8. The two graphs show that the potential speed up is realized in most cases.

The differences between the two quantities are shown clearly in Table 7-2, which gives the number of loops meeting the lower bound, and in Figure 7-9, which shows a lower bound on the efficiency of those that do not.

Two main conclusions can be drawn from these measurements: the potential speed up of software pipelining on innermost loops is high, and that this potential is achieved in most cases. Overall, at least 85 out of 117 loops are scheduled with optimal throughput. Since we can only establish a lower bound on the initiation interval, we can measure only the lower bound of the efficiency of the algorithm on the remaining loops. As shown in Figure 7-9, the measured efficiency for these loops is rather high.

For loops with neither non-trivial connected components nor con-

(a)

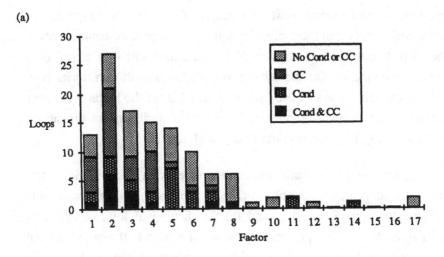

(b)

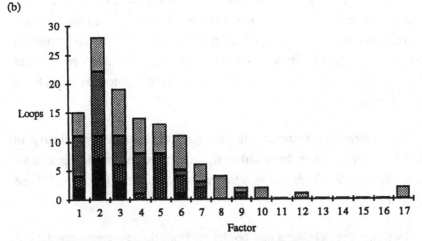

Figure 7-8: Effectiveness of software pipelining:
(a) potential and (b) achieved speed up

ditional statements, the potential speed up is spread mostly within the
range from a factor of 2 to 6. Such loops are simplest to handle and the
results are excellent: a large percentage of the loops meets the lower
bound and the efficiency of the remaining loops is very high. The percentage of loops meeting the lower bound decreases if they contain non-
trivial connected components or conditional statements. There are two
reasons, as further explained below: the lower bound is not as exact and
the heuristics do not perform as well.

Program	Number	Out of	%
No Cond or CC	45	48	94
CC	22	32	69
Cond	14	24	58
Cond & CC	4	13	31
Total	85	117	73

Table 7-2: Loops of 100% efficiency

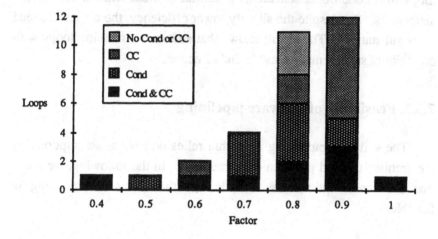

Figure 7-9: Lower bound of efficiency of remaining loops

Loops with non-trivial connected components generally have lower potential for speed up, as can be observed from Figure 7-8. The cycles in the precedence constraint graph tend to raise the initiation interval, and thus lower the potential speed up. Nodes in a connected components are more constrained in their schedule; the resulting resource conflicts are not reflected in the lower bound of the initiation interval. Also, nodes representing connected components are harder to accommodate; their scheduling constraints represent the union of their components and are therefore more stringent. These two factors lead to a lower percentage of programs meeting the lower bound. Nonetheless, the effect of software pipelining is still significant on these loops.

Before software pipelining is applied, conditional statements are first scheduled as compactly as possible to avoid code explosion. If the branches are longer than the calculated lower bound on the initiation interval, then it is possible that the schedule of the nodes in the condition statement has already violated the resource constraints for some initiation interval values above the lower bound. Therefore, the lower bound obtained is less strict than those of other loops. The potential speed up of loops with conditional statements is similar to those without conditional statements. So, despite the slightly lower efficiency, the achieved speed up is substantial. This result shows that software pipelining loops with conditional statements is feasible and effective.

7.3.3. Feasibility of software pipelining

The software pipelining algorithm relies on several assumptions on the architecture and program characteristics. In the following, we show that these assumptions are realistic and that software pipelining is feasible.

Software pipelining iterates with increasing target initiation intervals until a schedule is found. This iterative approach is feasible because, as shown in Table 7-2, a schedule can often be found in the first attempt. Even for loops not meeting the lower bound, a schedule can typically be found within a few iterations.

In software pipelining, the object code is sometimes unrolled to implement modulo variable expansion. Object code unrolling increases the size of the object code and increases the demand of registers. Figure 7-10 shows the distribution of the degrees of unrolling in the sample loops. A degree of one means that the loop is not unrolled at all. We observe that while unrolling is used about a third of the time, the degree of unrolling, fortunately, remains small.

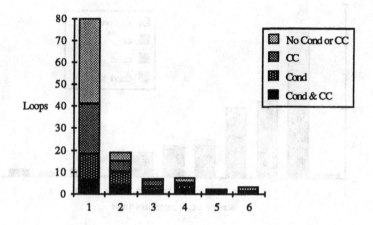

Figure 7-10: Distribution of the degrees of unrolling

In this software pipelining algorithm, register assignment is performed after the scheduling phase, and the scheduler assumes that enough registers are available. There are two 31-word register files, known as the A and M register file, in the Warp data path. They are used exclusively for storing operands for the floating-point adder and multiplier, respectively. All operands for the arithmetic units must first be stored in the register files before they can be used. Intermediate expressions are stored only in these register files, and are not written back into the data memory. In addition, registers are allocated to scalar variables in innermost loops for the duration of the loop.

The resource usage of these register files, shown in Figure 7-11, reflects the usage of the floating-point units: the utilization of the A register file is much greater than that of the M register file. In general, the assumption that there are enough registers is valid. The optimization of scheduling the graph in a top-down manner, as discussed in Section 4.3.3, reduces the lifetimes of the register values, and consequently, the register requirement. It is sometimes necessary to limit the number of registers allocated to scalar variables in the loop. And in several cases, described below, the loops cannot be software pipelined because of lack of registers.

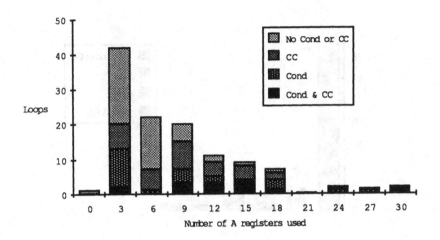

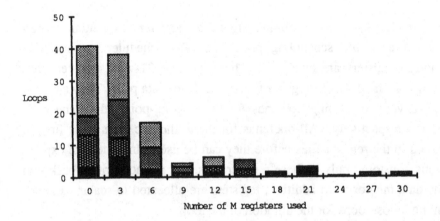

Figure 7-11: Register usage

Of the 122 loops, there are five that are not pipelined. All five loops have the same characteristics: they have unusually large loop bodies, produced by hand unrolling the loops in the source program. (Second level pipelining is preferable to source code unrolling.) Their characteristics are summarized as follows:

- Loop contains 17 connected components and 8 conditional statements.

Execution time of one iteration: 101
Minimum initiation interval: 87
Efficiency (lower bound):   86%

- Loop contains 1 connected component, which contains 60 components, 5 of which are conditional statements.

Execution time of one iteration: 142
Minimum initiation interval: 113
Efficiency (lower bound):   80%

- Loop contains 2 connected components, one of which has 13 components, 6 of which are conditional statements.

Execution time of one iteration: 72
Minimum initiation interval: 40
Efficiency (lower bound):   56%

- Loop contains 48 connected components and 8 conditional statements. A schedule is found for the initiation interval of 189 instructions, with degree of unrolling 2.

Execution time of one iteration: 207
Minimum initiation interval: 155
Efficiency (lower bound):   75%

- Loop contains 9 conditional statements and no non-trivial connected components. A schedule is found for the initiation interval of 134 instructions, with degree of unrolling 3.

Execution time of one iteration: 240
Minimum initiation interval: 132
Efficiency (lower bound):   55%

While the large numbers of operations in the loops make software pipelining difficult, they also make the technique unnecessary since the efficiency of using only hierarchical reduction is reasonably high. For the first three loops, the scheduler simply cannot find a schedule for an initiation interval shorter than the execution time of an iteration of the unpipelined loop. These results are compatible with those reported in Section 7.3.2, given that large numbers of connected components and conditional statements are present in the loops. For the last two loops, although a pipelined schedule can be found, it is not used because of the lack of registers. Both loops contain an extraordinarily large number of objects to be scheduled, and the potential gain of software pipelining is low. In the current implementation, if the execution time of an unpipelined iteration is above a certain threshold, software pipelining is not even attempted to cut down on compilation time.

7.4. Livermore Loops

The study of the large set of user programs provides a general picture of the effectiveness of applications on the machine. However, because of the large numbers of programs used, we can only give the statistics as a whole; we cannot list the code for each program, nor can we correlate the achieved performance with the individual loop characteristics. To compensate for this, we measured the performance of a single PC Warp cell on Livermore Loops. We chose the Livermore Loops because they are standard benchmarks and performance numbers have been reported for many different compilers and machines.

The performance of Livermore Loops [43] on a single Warp cell is shown in Table 7-3. The Livermore Loops were manually translated from FORTRAN to W2. The translation into W2 was straightforward, preserving loop structures and changing only syntax, except for kernels 15 and 16, which were translated from Feo's restructured loops [16]. The INVERSE and SQRT functions expanded into 7 and 19 floating-point operations, respectively. The EXP function in kernel 22 expanded into a calculation containing 19 conditional statements. The large numbers of conditional statements made the loop not pipelinable. In fact, the scheduler did not even attempt to pipeline this loop because the length of the loop (331 instructions) was beyond the threshold that it used to decide if pipelining was feasible. Kernels 16 and 20 were also not pipelined, because the calculated lower bound on the initiation interval were within 99% of the length of the unpipelined loop.

The second column contains lower bound figures on the optimality of the software pipelining technique. As they were obtained by dividing

Kernel	Optimality (lower bound)	Speed up	MFLOPS
1	1.00	8.25	6.2
2*	0.75	3.25	2.5
3	1.00	2.71	1.4
4*	1.00	2.71	1.4
5	0.94	1.12	0.6
6*	1.00	2.86	1.4
7	1.00	6.00	7.9
8*	1.00	2.29	8.2
9	1.00	4.27	7.7
10	0.85	5.31	3.4
11	0.90	1.30	0.5
12	1.00	4.00	1.7
13	1.00	2.63	2.0
14**	1.00	3.32	2.2
15	0.85	5.50	5.9
16	1.00	1.00	0.3
17	1.00	1.20	0.8
18	0.97	3.70	7.5
19	1.00	1.24	0.7
20	0.99	1.00	0.9
21	1.00	6.00	3.0
22***	0.56	1.00	1.1
23	1.00	1.10	1.1
24	1.00	1.33	0.4
H-Mean			1.2

*   Compiler directives to disambiguate array references used

**   Multiple loops were merged into one

*** EXP function expanded into 19 IF statements

Table 7-3: Performance of Livermore Loops

the lower bound on the initiation interval by the achieved interval value, they represent a lower bound on the achieved efficiency. For kernels containing multiple loops, the figure given is the mean calculated by weighing each loop by its execution time. The performance of the Livermore Loops is consistent with that of the users' programs. Except for kernel 22, which has an extraordinary amount of conditional branching due to the particular EXP library function, near-optimal, and often times, optimal code is obtained. The speed up factors in the third column are the ratios of the execution time between an unpipelined and a pipelined kernel. All branches of conditional statements were assumed to be taken half the time. The MFLOPS rates given in the last column are for single-precision floating-point arithmetic.

The figures in the second column indicate that optimal efficiency is achieved for most of the Livermore Loops. Although consistently good results are obtained, the speed up fluctuates dramatically. The achieved MFLOPS rates range from 0.4 to 8.2 MFLOPS. First, we note that there is very little correlation between the optimality and the achieved execution rates. The MFLOPS rate of the kernels 3 to 6 are very low, but their optimality figures are typical and are reasonably high. On the other hand, kernels 7 to 9 have high MFLOPS rates and are 100% optimal. There is a high potential of parallelism in these loops, and the scheduler is able to exploit it fully.

In general, the performance of the generated code depends mainly on the characteristics of the program. If there is parallelism within and across iterations of a loop, the scheduler is able to find it and exploit it. As discussed in Section 7.2.2, parallelism is limited by data dependency and the distribution of resource usage:

1. *Data dependency.* Consider kernel 5 of the Livermore Loops:

```
FOR i := 0 TO n-1 DO BEGIN
    X[i] := Z[i]*(Y[i]-X[i-1]);
END;
```

The multiplications and additions are serialized because of the data dependencies. Even if the adder and the multiplier were not pipelined, only one unit can be busy at any one time. But since the multiplier and the adder are seven stage pipelined, each iteration takes 14 clocks. Just by this consideration alone, Warp is limited to a peak performance of 0.7 MFLOPS on this loop.

Inter-iteration data dependency, or recurrences, do not necessarily mean that the code is serialized. This is one important advantage that VLIW architectures have over vector machines. As long as there are other operations that can execute in parallel with the serial computation, a high computation rate can still be obtained.

2. *Critical resource bottleneck.* The execution speed of a program is limited by the most heavily used resource. Unless *both* the floating-point multiplier and adder are the most critical resources, the peak MFLOPS rate cannot be achieved. Programs containing no multiplications cannot run faster than 5 MFLOPS since the multiplier is idle all the time. For example, since the integer unit is the most heavily used resource in kernel 13, the MFLOPS measure is naturally low.

7.5. Summary and discussion

Many systolic array programs have been developed for the Warp prototype machine, despite the restrictions of compile-time flow control and homogeneous programming. This is not surprising because all previously proposed systolic algorithms in the literature belong to this restricted class of computation. All the programs in the sample have unidirectional data flow, and the compiler is able to generate code that fully utilizes the array of cells. Once a cell is activated, after possibly a small delay, it never stalls waiting for data. By rearranging I/O operations at code scheduling time, the code generated for a cell is just as efficient as it would have been if the cell did not have to interact with other cells.

At the cell level, our analysis shows that the cell code scheduler performs well for the sample of user programs in the experiment. More precisely, an average of a three-fold increase in speed has been observed, and, on average, the efficiency of the scheduler is shown to be at least 20% within optimal. Software pipelining is effective; optimal throughput is achieved for at least 73% of the loops. Hierarchical reduction is also important: It makes software pipelining applicable to loops with conditional statements, and allows consistent speed up be achieved for all programs.

Although an average of 28% utilization is a respectable figure for high-performance processors, there seems a large discrepancy between the achieved performance and the measured efficiency of the scheduler. The analysis in Section 7.2.2 provides an answer to this question. The major reason is that the high peak rate of the Warp array is achieved through parallel and pipelined functional units. As the execution time is limited by the most heavily used resource, the effective computation rate of the machine is generally much lower than the peak. For the sample of programs studied, if the critical resource in every cell is busy at all times, the average performance obtained would still only be 50 MFLOPS. This number can be improved only by increasing the bandwidth of the functional units in the data path. Modifications, such as reducing the number of pipelined stages in the arithmetic units, do not change this figure.

The analysis also pinpoints one potential area for improvement. The computation and the I/O units are not used in parallel in many of the programs where data partitioning model of computation is used. This model is extremely powerful and should be supported effectively. Overlapping the computation and I/O phases is difficult because they are two unrelated tasks with different control flow. In many cases, it is possible to apply high-level transformations to merge the different phases in software. However, a general solution to this problem is to provide independent sequencing control for the I/O process. This support is provided in iWarp.

8
Conclusions

The Warp architecture represents a significant breakthrough in systolic computing. Systolic arrays have previously been known to be highly efficient, but special-purpose hardware architectures. The cost of dedicated hardware is so prohibitive that although many different systolic algorithms have been designed, few have been implemented. The Warp project demonstrates that programmability can be achieved without sacrificing efficiency. Even in its prototype form, the Warp machine can already deliver execution speeds for many numerical oriented applications that rival existing supercomputers. The availability of the single-chip iWarp processor will make a significant impact on the practice of parallel computing. Arrays of thousands of cells are feasible, programmable, and much cheaper than many other supercomputers of comparable power.

This book studies one of the vital aspects of the Warp machine: programmability. My thesis is that high-performance systolic arrays can be programmed effectively using a high-level language. The ideas and techniques in the book have been validated by the W2 compiler implementation, which has been used extensively by a large user community. Optimal performance is obtained for many simple, classical systolic programs, such as convolution and matrix multiplication. The availability of the language and the compiler have made possible the development and implementation of many new, complex systolic algorithms. The compiler provided the programmability and efficiency essential to making Warp a valuable resource in computation intensive domains. This research has extended the domain of systolic processing from simple mathematical recurrences in custom VLSI implementations to arbitrarily complex programs on powerful and programmable processors.

Specifically, this work makes contributions to two major research areas: a machine abstraction and compiler optimizations for systolic arrays, and code scheduling techniques for horizontally microcoded processors.

8.1. Machine abstraction for systolic arrays

We propose to model a high-performance systolic array as an array of simple sequential processors with asynchronous communication. While this machine model is well-known in general computing, its applicability to systolic arrays is not apparent. This is because systolic algorithm designers typically exercise extreme control over the timing of the communication between cells to guarantee a high utilization of the cells. This work shows that, in fact, for the very reason of efficiency, we should select the asynchronous communication model. The internal timing of a parallel and pipelined processor interacts with the array level of parallelism, and this timing must be considered in generating efficient code for the array. Therefore, if the internal parallelism of the cells in an array were to be hidden from the user by the use of a high-level programming language, then the timing of the interaction between cells would also have to be under the jurisdiction of the compiler.

Efficiency typical of systolic array algorithms can be obtained for unidirectional systolic algorithms written under this asynchronous communication model. The high-level semantics of asynchronous communication avoids over-specification in the interaction between cells. Simple analysis can be performed independently on each cell program to relax the original ordering of communication operations in the sequential program. This allows the scheduler to rearrange the communication operations to use the internal parallelism effectively. The sequencing constraints between communication operations are modeled in a manner similar to data dependency constraints between computational operations. This correspondence of systolic array specific constraints to well known concepts in code scheduling allows us to apply established tools and algorithms to our specific problem.

Our machine abstraction hides two low-level issues: cell implementation details and synchronization between cells. This abstraction represents a significant data point in the space of machine models for systolic arrays. First, it offers both efficiency and generality. Our experience in using the model and the compiler supporting the model indicates that there is no need for lower level tools. This machine abstraction can be used as a target machine model for higher level programming tools. It is likely that different higher level tools will be developed to capture characteristics of different array usage models. Our abstraction may serve as a common base model for all these different approaches. Second, the machine abstraction can be used for a wide range of hardware implementations. Although the asynchronous communication model is used in the abstraction, it is not necessary that this communication model be supported directly in hardware. This is important because many systolic algorithms, including all previously published ones, do not need dynamic flow control. We have shown that this machine abstraction is recommended even for machines without the dynamic flow control hardware because it is amenable to compiler optimizations.

8.2. Code scheduling techniques

This research work shows that software pipelining is a *practical*, *efficient*, and *general* technique for scheduling the parallelism in a horizontally microcoded or VLIW machine. Previous implementations of software pipelining either depend on specialized hardware, or limit the kind of loops to which it can be applied. This research develops software pipelining into a complete algorithm that is based on software heuristics. In particular, the algorithm for cyclic program graphs has been significantly improved. Software pipelining of innermost loops is the primary reason for the high attainable performance on Warp.

This work also proposes a unified approach to scheduling both within and across basic blocks called hierarchical reduction. By modeling an entire construct in a manner similar to an operation in a basic block, all scheduling techniques previously applicable only within a basic block are now applicable to all constructs. The important effects on execution time are: loops with conditional statements can be pipelined, innermost loops containing large numbers of blocks and operations can be effectively compacted, the prologs and epilogs of innermost loops can be overlapped with scalar operations outside the loops. Hierarchical reduction is important in ensuring that a consistent performance improvement can be obtained for all programs.

The algorithms of software pipelining and hierarchical reduction are relatively simple, and can be implemented without undue effort. Comparison with the small sample of painstakingly hand-microcoded applications indicates that the generated code is at least as efficient. The experimental data on a large set of programs shows that optimal throughput can be obtained for the innermost loops of many of the applications running on the Warp machine.

The study of the performance of the compiler and application programs on Warp leads to several observations on architecture design.

First, the effective use of the highly parallel and pipelined data paths by compiler-generated code suggests a new alternative to vector processing. Custom generation of microcode is much more flexible and adaptive to the user's program. It alleviates the need for intermediate buffering of vectors of data. Moreover, efficient code can be generated for programs that are hard to vectorize. Software pipelining can support general recurrences in a loop, as well as conditional statements. Lastly, scalar and iterative constructs can be easily intermixed, thus significantly reducing the penalty associated with short vectors.

Second, the systolic architecture is a feasible and effective parallel machine organization. Systolic arrays possess two fundamental attractive attributes: their scalable architecture offers a solution to meet the ever-growing demands on computation speed, and their fast local interconnection between cells uniquely supports algorithms of fine-grain parallelism effectively. The Warp machine demonstrates that a programmable systolic array of high-performance processors is feasible, and this research shows that the power of a systolic array can be harnessed without undue programming efforts. We have laid a foundation for further research in both systolic architectures and higher level programming tools. With the development of more sophisticated hardware and software support, systolic arrays will develop into a major resource for computation intensive applications.

References

[1]   Aho, A. V., Sethi, R., and Ullman J. D.
        Compilers.
        Addison-Wesley, 1986.

[2]   Aiken, A. and Nicolau, A.
        Perfect Pipelining: A New Loop Parallelization Technique.
        Technical Report, Cornell University, Oct., 1987.

[3]   Annaratone, M., Bitz, F., Clune E., Kung H. T., Maulik, P.,
        Ribas, H., Tseng, P., and Webb, J.
        Applications of Warp.
        In *Proc. Compcon Spring 87*, pages 272-275. IEEE Computer
            Society, San Francisco, Feb., 1987.

[4]   Annaratone, M., Bitz, F., Deutch, J., Hamey, L., Kung, H. T.,
        Maulik, P. C., Tseng, P., and Webb, J. A.
        Applications Experience on Warp.
        In *Proc. 1987 National Computer Conference*, pages 149-158.
            AFIPS, Chicago, June, 1987.

[5]   Brent, R. P. and Kung, H. T.
        Systolic VLSI Arrays for Polynomial GCD Computation.
        IEEE Transactions on Computers C-33(8):731-736, August,
            1984.

[6]   Chen, M.
        Space-Time Algorithms: Semantics and Methodology.
        Technical Report 5090:TR:83, California Institute of Technology,
            May, 1983.

[7]   Chen, M.
        A Parallel Language and Its Compilation to Multiprocessor
            Machines or VLSI.
        In *Proc. 13th Annual ACM Symposium on Principles of Program-
            ming Languages*. Jan., 1986.

[8]   Coffman, E. G., Jr.
        Computer and Job-shop Scheduling Theory.
        John Wiley & Sons, 1976.

[9]   Colwell, R. P., Nix, R. P., O'Donnell, J. J., Papworth, D. B., and
        Rodman, P. K. .
        A VLIW Architecture for a Trace Scheduling Compiler.
        In *Proc. Second Intl. Conf. on Architectural Support for Pro-
             gramming Languages and Operating Systems*, pages 180-192.
             Oct., 1987.

[10]   Cydrome Inc.
        CYDRA 5 Directed Dataflow Architecture.
        1987.

[11]   Dantzig, G. B., Blattner, W. O. and Rao, M. R.
        All Shortest Routes from a Fixed Origin in a Graph.
        In *Theory of Graphs*, pages 85-90. Rome, July, 1967.

[12]   Dasgupta, S.
        The Organization of Microprogram Stores.
        Comput. Surv. 11(1):39-65, March, 1979.

[13]   Delosme, J.-M., Ipsen, I. C. F.
        Design Methodology for Systolic Arrays.
        In *Proc. SPIE Symp.*, pages 245-259. 1986.

[14]   Ebcioglu, K.
        A Compilation Technique for Software Pipelining of Loops with
             Conditional Jumps.
        In *Proc. 20th Annual Workshop on Microprogramming*. Dec.,
             1987.

[15]   Ellis, J. R.
        Bulldog: A Compiler for VLIW Architectures.
        PhD thesis, Yale University, 1985.

[16]   Feo, J. T.
        An Analysis of the Computational and Parallel Complexity of the
             Livermore Loops.
        Parallel Computing 7(2):163-186, June, 1988.

[17]   Fisher, J. A.
        *The Optimization of Horizontal Microcode Within and Beyond
             Basic Blocks: An Application of Processor Scheduling with
             Resources.*
        PhD thesis, New York Univ., Oct., 1979.

[18]   Fisher, J. A.
        2^n-Way Jump Microinstruction Hardware and an Effective In-
            struction Binding Method.
        In *Proc. 13th Annual Workshop on Microprogramming*, pages
            64-75. November, 1980.

[19]   Fisher, J. A.
        Trace Scheduling: A Technique for Global Microcode Compac-
            tion.
        IEEE Transactions on Computers C-30(7):478-490, July, 1981.

[20]   Fisher, J. A.
        Very Long Instruction Word Architectures and the ELI-512.
        In *Proc. Tenth Annual Symposium on Computer Architecture*,
            pages 140 - 150. Stockholm, June, 1983.

[21]   Fisher, J. A., Ellis, J. R., Ruttenberg, J. C. and Nicolau, A.
        Parallel Processing: A Smart Compiler and a Dumb Machine.
        In *Proc. ACM SIGPLAN '84 Symp. on Compiler Construction*,
            pages 37-47. Montreal, Canada, June, 1984.

[22]   Fisher, A. L., Kung, H. T. and Sarocky, K.
        Experience with the CMU Programmable Systolic Chip.
        In *Proceedings of SPIE Symposium, Vol. 495, Real-Time Signal
            Processing VII*. Society of Photo-Optical Instrumentation En-
            gineers, August, 1984.

[23]   Fisher, J. A., Landskov, D. and Shriver, B. D.
        Microcode Compaction: Looking Backward and Looking For-
            ward.
        In *Proc. 1981 National Computer Conference*, pages 95-102.
            1981.

[24]   Fisher, J. A. and O'Donnell, J. J.
        VLIW Machines: Multiprocessors We Can Actually Program.
        In *Proc. Compcon Spring 84*, pages 299-305. IEEE Computer
            Society, February, 1984.

[25]   Floyd, R. W.
        Algorithm 97: Shortest Path.
        Comm. ACM 5(6):345, 1962.

[26]   Fortes, J. A. B., Moldovan, D. I.
        Parallelism Detection and Transformation Techniques Useful for
            VLSI Algorithms.
        Journal of Parallel and Distributed Computing 2:277-301, 1985.

[27]   Garey, M. R. and Johnson, D. S.
        *Computers and Intractability A Guide to the Theory of
            NP-Completeness.*
        Freeman, 1979.

[28]   Gross, T. and Lam, M.
        Compilation for a High-performance Systolic Array.
        In *Proc. ACM SIGPLAN 86 Symposium on Compiler
            Construction*, pages 27-38. June, 1986.

[29]   Gross, T. R. and Hennessy, J. L.
        Optimizing Delayed Branches.
        In *Proc. 15th Annual Workshop on Microprogramming*, pages
            114-120. 1982.

[30]   Gross, T., Kung, H. T., Lam, M. and Webb, J.
        Warp as a Machine for Low-Level Vision.
        In *Proc. IEEE International Conference on Robotics and
            Automation*, pages 790-800. March, 1985.

[31]   Hsu, P.
        Highly Concurrent Scalar Processing.
        PhD thesis, University of Illinois at Urbana-Champaign, 1986.

[32]   Isoda, S., Kobayashi, Y., and Ishida, T.
        Global Compaction of Horizontal Microprograms Based on the
            Generalized Data Dependency Graph.
        IEEE Transactions on Computers c-32(10):922-933, October,
            1983.

[33]   Kuck, D. J., Kuhn, R. H., Padua, D. A., Leasure, B. and Wolfe,
        M.
        Dependence Graphs and Compiler Optimizations.
        In *Proc. ACM Symposium on Principles of Programming
            Languages*, pages 207-218. January, 1981.

[34]   Kung, H. T.
        Why Systolic Architectures?
        Computer Magazine 15(1):37-46, January, 1982.

[35]   Kung, H. T.
        Memory Requirements for Balanced Computer Architectures.
        Journal of Complexity 1(1):147-157, 1985.

[36]   Kung, H. T. and Lam, M.
        Wafer-Scale Integration and Two-Level Pipelined Implemen-
          tations of Systolic Arrays .
        Journal of Parallel and Distributed Computing 1(1):32-63, 1984.
        A preliminary version appears in Proc. Conference on Advanced
          Research in VLSI, MIT, January 1984, pp. 74-83.

[37]   Kung, H. T. and Webb, J.A.
        Global Operations on the CMU Warp Machine.
        In *Proc. 1985 AIAA Computers in Aerospace V Conference*,
          pages 209-218. American Institute of Aeronautics and
          Astronautics, Oct., 1985.

[38]   Kung, H. T. and Webb, J. A.
        Mapping Image Processing Operations onto a Linear Systolic
          Machine.
        Distributed Computing 1(4):246-257, 1986.

[39]   Lah, J. and Atkin, E.
        Tree Compaction of Microprograms.
        In *Proc. 16th Annual Workshop on Microprogramming*, pages
          23-33. Oct., 1982.

[40]   Lam, M. and Mostow, J.
        A Transformational Model of VLSI Systolic Design.
        Computer 18(2), Feb., 1985.
        An earlier version appears in Proc. 6th International Symposium
          on Computer Hardware Description Languages and their Ap-
          plications, May, 1983.

[41]   Leiserson, C. E. and Saxe, J. B.
        Optimizing Synchronous Systems.
        Journal of VLSI and Computer Systems 1(1):41-68, 1983.

[42]   Linn, J. L.
        SRDAG Compaction - A Generalization of Trace Scheduling to
          Increase the Use of Global Context Information.
        In *Proc. 16th Annual Workshop on Microprogramming*, pages
          11-22. 1983.

[43]   McMahon, F. H.
        Lawrence Livermore National Laboratory FORTRAN Kernels:
          MFLOPS.
        1983.

[44]   Patel, J. H. and Davidson, E. S.
        Improving the Throughput of a Pipeline by Insertion of Delays.
        In *Proc. 3rd Annual Symposium on Computer Architecture*, pages
            159-164. Jan., 1976.

[45]   Pomerleau, D. A., Gusciora, G. L., Touretzky, D. S. and Kung,
        H. T.
        Neural Network Simulation at Warp Speed: How We Got 17 Mil-
            lion Connections Per Second.
        In *Proceedings of 1988 IEEE International Conference on Neural
            Networks*, pages 143-150. July, 1988.

[46]   Quinton, P.
        Automatic Synthesis of Systolic Arrays from Uniform Recurrent
            Equations.
        In *Proc. 11th Annual Symposium on Computer Architecture*.
            1984.

[47]   Rau, B. R. and Glaeser, C. D.
        Some Scheduling Techniques and an Easily Schedulable
            Horizontal Architecture for High Performance Scientific
            Computing.
        In *Proc. 14th Annual Workshop on Microprogramming*, pages
            183-198. Oct., 1981.

[48]   Sejnowski, T. J., and Rosenberg, C. R.
        Parallel Networks that Learn to Pronounce English Text.
        Complex Systems 1(1):145-168, 1987.

[49]   Shustek, L. J.
        Analysis and Performance of Computer Instruction Sets.
        PhD thesis, Stanford University, May, 1977.

[50]   Electrotechnical Laboratory.
        *SPIDER (Subroutine Package for Image Data Enhancement and
            Recognition)*.
        Joint System Development Corp., Tokyo, Japan, 1983.

[51]   Su, B., Ding, S. and Jin, L.
        An Improvement of Trace Scheduling for Global Microcode
            Compaction.
        In *Proc. 17th Annual Workshop in Microprogramming*, pages
            78-85. Dec., 1984.

[52] Su, B., Ding, S., Wang, J. and Xia, J.
GURPR – A Method for Global Software Pipelining.
In *Proc. 20th Annual Workshop on Microprogramming*, pages
88-96. Dec., 1987.

[53] Su, B., Ding, S. and Xia, J.
URPR – An Extension of URCR for Software Pipeline.
In *Proc. 19th Annual Workshop on Microprogramming*, pages
104-108. Oct., 1986.

[54] Tarjan, R. E.
Depth first search and linear graph algorithms.
SIAM J. Computing 1(2):146-160, 1972.

[55] Tokoro, M., Tamura, E. and Takizuka, T.
Optimization of Microprograms.
IEEE Transactions on Computers c-30(7):491-504, July, 1981.

[56] Touzeau, R. F.
A Fortran Compiler for the FPS-164 Scientific Computer.
In *Proc. ACM SIGPLAN '84 Symp. on Compiler Construction*,
pages 48-57. June, 1984.

[57] Weiss, S. and Smith, J. E.
A Study of Scalar Compilation Techniques for Pipelined Super-
computers.
In *Proc. Second Intl. Conf. on Architectural Support for Pro-
gramming Languages and Operating Systems*, pages 105-109.
Oct., 1987.

[58] Wolfe, M. J.
Optimizing Supercompilers for Supercomputers.
PhD thesis, University of Illinois at Urbana-Champaign, October,
1982.

[59] Woo, B., Lin, L. and Ware, F.
A High-Speed 32 Bit IEEE Floating-Point Chip Set for Digital
Signal Processing.
In *Proc. 1984 IEEE International Conference on Acoustics,
Speech and Signal Processing*, pages 16.6.1-16.6.4. 1984.

[60] Wood, G.
Global Optimization of Microprograms Through Modular Control Constructs.
In *Proc. 12th Annual Workshop in Microprogramming*, pages
1-6. 1979.

[61]   Young, D.
        Iterative Solution of Large Linear Systems.
        Academic Press, New York, 1971.

Index

CONNECTIONIST APPROACHES TO LANGUAGE LEARNING

edited by

DAVID TOURETZKY

Carnegie Mellon University

A Special Issue of Machine Learning on Connectionist Approaches to Language Learning

Reprinted from Machine Learning
Vol. 7, Nos. 2–3 (1991)

KLUWER ACADEMIC PUBLISHERS
BOSTON/DORDRECHT/LONDON

THE KLUWER INTERNATIONAL SERIES
IN ENGINEERING AND COMPUTER SCIENCE

KNOWLEDGE REPRESENTATION, LEARNING AND EXPERT SYSTEMS

Consulting Editor

Tom Mitchell
Carnegie Mellon University

MACHINE LEARNING/*Vol. 7   Nos. 2/3 September 1991*

CONNECTIONIST APPROACHES TO LANGUAGE LEARNING
A Special Issue of Machine Learning on
Connectionist Approaches to Language Learning

Distributors for North America:
Kluwer Academic Publishers
101 Philip Drive
Assinippi Park
Norwell, Massachusetts 02061 USA

Distributors for all others countries:
Kluwer Academic Publishers Group
Distribution Centre
Post Office Box 322
3300 AH Dordrecht, THE NETHERLANDS

Library of Congress Cataloging-in-Publication Data

Connectionist approaches to language learning / [edited] by David
 Touretzky.
     p.   cm. — (The Kluwer international series in engineering and
 computer science. Knowledge representation, learning, and expert
 systems)
   "Also been published as a special issue of Machine learning,
 volume 7, issue 2/3"—P.
   Includes index.
   ISBN 0-7923-9216-7 (acid-free paper)
   1. Machine learning.   2. Language and languages—Study and
 teaching.   I. Touretzky, David S.   II. Series.
 Q325.5.C67   1991
 006.3′1–dc20   91-25064
                                                                 CIP

Printed on acid-free paper.

Printed in the United States of America

Machine Learning 7, 105 – 107 (1991)
© 1991 Kluwer Academic Publishers, Boston. Manufactured in The Netherlands.

Introduction

The goal of this special issue is twofold: first, to acquaint members of the machine learning community with the latest results in connectionist language learning, and second, to
make these five inter-related papers available in a single publication as a resource for others
working in the area. In the remainder of this introduction I will sketch what it is that I
think the connectionist approach offers us, and how the papers in this special issue advance
the state of the art. But this is not going to be a cheerleading piece about the wonders
of "brain-style computation" and the imminent death of symbolic AI. Rather, I hope to
tempt the reader into examining some novel ideas that expand the current scope of AI.

"Connectionism" is not a single distinct approach to learning. It is a collection of loosely-
related ideas covering a broad range of intellectual territory: dynamical systems theory,
computational neuroscience, cognitive modeling, and so on. While the ultimate goal of
the connectionist enterprise is supposed to be discovering how brains embody intelligence,
current connectionist learning models are nowhere near achieving either general intelligence
or neural plausibility. (AI as a whole is still far from producing generally intelligent agents,
so this is not an indictment of any one particular approach.)

Despite the fact that today's "neural net" models are not brain-like to any meaningful
extent, they have interesting insights to offer the Machine Learning community. The connectionist approach to computation is quite different from discrete symbol manipulation. In
some cases, such as the paper by Mozer and Bachrach, this difference manifests itself in
useful behavioral properties and broader/alternative conceptualizations of the learning problem. In other cases, such as the paper by Servan-Schreiber, Cleeremans, and McClelland,
and the following one by Elman, it has led to some surprising suggestions about human
cognition. The papers in this special issue represent some of the best connectionist work
to date on the problem of language learning. They span the range from solid theoretical
analysis (Porat and Feldman) to bold conjecture (Pollack). Together they demonstrate the
richness of this exciting multi-disciplinary movement.

So what exactly is the connectionist approach to computation? Common themes have
been massive parallelism and simple computing units with limited communication. Beyond
these general properties, connectionist models can be divided into two major classes. In
localist models, individual units represent specific items or concepts. Localist models are
generally constructed by hand, or in the case of Porat and Feldman's work, by a discrete
learning algorithm. The principal advantages of localist models are that they are easy both
to construct and to analyze.

Distributed models represent information as diffuse patterns of activity of a collection
of units, and are usually constructed by gradient-descent learning algorithms such as back-
propagation. The principal advantages of these models are that they develop their own representations *de novo*, and naturally generalize to novel inputs. A disadvantage is that their
behavior is harder to control; typically many iterations of the learning algorithm are required
to achieve acceptable performance on the training set, and generalization ability must be
tested empirically. A special case of distributed models, the Simple Recurrent Network

(SRN), can process sequences of symbols by inputting them one at a time. Recurrent or "feedback" connections allow the SRN to incrementally construct representations for sequences, or whichever bits of them are relevant, as locations in a high-dimensional vector space.

In the first paper in this special issue, Porat and Feldman consider the tractability of learning a regular language (or equivalently, a finite state automaton) when input examples are lexicographically ordered. Although the construction of a minimum-state deterministic FSA is known to be NP-hard, they show that the problem can be solved in polynomial time, with limited memory (no storage of previously-observed training examples), when lexicographic ordering is assumed. At the conclusion of the paper they describe an implementation of their algorithm by a localist-type connectionist network.

Mozer and Bachrach are also concerned with the efficient induction of finite state automata. They consider the problem of a robot wandering around in a simulated finite-state environment that must learn to predict the effects of its actions. A symbolic learning algorithm by Schapire and Rivest solves this problem using a structure called an *update graph*. Mozer and Bachrach show that a connectionist implementation of the update graph idea, based on backpropagation learning, learns small worlds more quickly than the original algorithm. Furthermore, the discrete symbolic update graph structure turns out to be merely a limiting case of a more general, continuous representation created by backprop through gradient-descent learning.

Approximation of discrete, finite-state phenomena by continuous dynamical systems is a topic of much current research. The next three papers in this special issue analyze the internal representations created by various recurrent networks. Servan-Schreiber, Cleeremans, and McClelland train an Elman-style Simple Recurrent Network on strings from a moderately complex regular language. They use hierarchical clustering to analyze the hidden unit activation patterns, revealing a complex state structure that is richer than the minimal-state DFSA describing the same data. Perhaps their most striking result is that, for certain problems involving one FSA embedded in two places inside another, the resulting grammar is not learnable due to the difficulty of keeping the two embeddings separate. However, if the arc transition probabilities are altered from the usual .5/.5 to, say, .6/.4 for one embedding and .4/.6 for the other, this statistical marker distinguishes the two sufficiently so that the grammar can be learned. The suggestion Servan-Schreiber, Cleeremans, and McClelland make is that statistical variation may also be an important cue in human language learning. For example, although English permits arbitrary embedded clauses, in normal conversation the distribution of syntactic structures for embedded clauses is different from the distribution of structures for surface clauses. This might be a source of unconscious cues to help language learners master embedding.

Elman's paper provides additional insight into embedding phenomena. He considers the problem of learning context-free rather than regular grammars from examples. Recognizing context-free languages requires some sort of stack. The Simple Recurrent Network can learn to simulate a limited-depth stack in the structure of its hidden unit states. What's particularly interesting is that the automaton constructed by the SRN generalizes to sentences that involve slightly more complex structures than the ones in the training set. One cannot hope to get this sort of generalization from a purely symbolic FSA induction algorithm, since it results from the existence of additional states, not required by the training set, that arise automatically as a result of the recursive structure of the task and the continuous nature of the SRN's state space. Elman also introduces a new graphical technique for studying network behavior based on principal components analysis. He shows that sentences with multiple levels of embedding produce state space trajectories with an intriguing self-similar structure.

The development and shape of a recurrent network's state space is the subject of Pollack's paper, the most provocative in this collection. Pollack looks more closely at a connectionist network as a continuous dynamical system. He describes a new type of machine learning phenomenon: induction by phase transition. He then shows that under certain conditions, the state space created by these machines can have a fractal or chaotic structure, with a potentially infinite number of states. This is graphically illustrated using a higher-order recurrent network trained to recognize various regular languages over binary strings. Finally, Pollack suggests that it might be possible to exploit the fractal dynamics of these systems to achieve a generative capacity beyond that of finite-state machines.

It remains to be seen whether dynamical systems theory will permanently alter our understanding of symbolic phenomena, or merely provide an interesting diversion from classical automata theory. Our understanding of symbol processing in recurrent connectionist networks is still at an early, almost pretheoretic stage. Conjectures about how an infinite state space relates to Turing machines—or to real neural representations—are admittedly premature at this point, but they are tantalizing nonetheless. By studying these papers, the reader can share in the excitement of working at the connectionist frontier.

David S. Touretzky
School of Computer Science
Carnegie Mellon University
Pittsburgh, PA 15213-3890

Machine Learning, 7, 109–138 (1991)
© 1991 Kluwer Academic Publishers, Boston. Manufactured in The Netherlands.

Learning Automata from Ordered Examples

SARA PORAT*
Department of Computer Science, University of Rochester

JEROME A. FELDMAN[†]
Department of Computer Science, University of Rochester

Abstract. Connectionist learning models have had considerable empirical success, but it is hard to characterize exactly what they learn. The learning of finite-state languages (FSL) from example strings is a domain which has been extensively studied and might provide an opportunity to help understand connectionist learning. A major problem is that traditional FSL learning assumes the storage of all examples and thus violates connectionist principles. This paper presents a provably correct algorithm for inferring any minimum-state deterministic finite-state automata (FSA) from a complete ordered sample using limited total storage and without storing example strings. The algorithm is an iterative strategy that uses at each stage a current encoding of the data considered so far, and one single sample string. One of the crucial advantages of our algorithm is that the total amount of space used in the course of learning for encoding any finite prefix of the sample is polynomial in the size of the inferred minimum state deterministic FSA. The algorithm is also relatively efficient in time and has been implemented. More importantly, there is a connectionist version of the algorithm that preserves these properties. The connectionist version requires much more structure than the usual models and has been implemented using the Rochester Connectionist Simulator. We also show that no machine with finite working storage can iteratively identify the FSL from arbitrary presentations.

Keywords. Learning, finite automata, connectionist

1. Introduction

The ability to adapt and learn has always been considered a hallmark of intelligence, but machine learning has proved to be very difficult to study. There is currently a renewed interest in learning in the theoretical computer science community (Valiant, 1984; Valiant, 1985; Kearns, et al., 1987; Natarajan, 1987; Rivest & Schapire, 1987; Rivest & Schapire, 1987) and a, largely separate, explosive growth in the study of learning in connectionist networks (Hinton, 1987). One purpose of this paper is to establish some connections (sic) between these two research programs.

The setting for this paper is the abstract problem of inferring Finite State Automata (FSA) from sample input strings, labelled as + or − depending on whether they are to be accepted or rejected by the resulting FSA. This problem has a long history in theoretical learning studies (Angluin, 1976; Angluin, 1981; Angluin, 1987) and can be easily mapped to common connectionist situations. There are arguments (Brooks, 1987) that interacting FSA constitute a natural substrate for intelligent systems, but that issue is beyond the scope of this paper.

---

*Current Address: Science & Technology, IBM Israel Ltd., Technion City, Haifa, Israel.
[†]Current Address: International Computer Science Institute, Berkeley, CA.

5

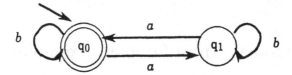

Figure 1. A parity FSA.

We will start with a very simple sample problem. Suppose we would like a learning machine to compute an FSA that will accept those strings over the alphabet $\{a, b\}$ that contain an even number of a's. One minimal answer would be the following two-state FSA shown in Figure 1.

We adopt the convention that states drawn with one circle are rejecting states and those drawn with a double circle are accepting. The FSA always starts in state q_0, which is accepting iff the empty string λ is to be accepted. We will present in Section 3 an algorithm that will always learn the minimum state deterministic FSA for any finite state language which is presented to the learning algorithm in strict lexicographic order. There are a number of issues concerning this algorithm, its proof and its complexity analysis that are independent of any relation to parallel and connectionist computation.

It turns out that the "even a's" language is the same as the well-studied "parity problem" in connectionist learning (Hinton, 1987). The goal there is to train a network of simple units to accept exactly binary strings with an even number of 1's. In the usual connectionist situation, the entire string (of fixed length) is presented to a bottom layer of units and the answer read from a pair of decision units that comprise the top layer. There are also intermediate (hidden) units and it is the weights on connections among all the units which the connectionist network modifies in learning.

The parity problem is very difficult for existing connectionist learning networks and it is instructive to see why this is so. The basic reason is that the parity of a string is a strictly global property and that standard connectionist learning techniques use only local weight-change rules. Even when a network can be made to do a fairly good job on a fixed-length parity problem, it totally fails to generalize to shorter strings. Of course, people are also unable to compute the parity of a long binary string in parallel. What we do in this situation is much more like the FSA of Figure 1. So one question concerns the feasibility of connectionist FSA systems.

There are many ways to make a connectionist version of an FSA like that of Figure 1. One of the simplest assigns a connectionist unit to each state and to the answer units $+$ and $-$. It is convenient to add an explicit termination symbol $\vdash$ and to use conjunctive connections (Feldman & Ballard, 1982) to capture transitions. The "current input letter" is captured as the activity of exactly one of the top three units. Figure 2 is the equivalent of Figure 1 under this transformation.

Thus unit 0 corresponds to the accepting state q_0 in Figure 1 because when it is active and the input symbol is $\vdash$, the answer $+$ is activated. Similarly, activity in unit 1 and in the unit for a leads to activity in unit 0 for the next time step. Note that activity is allowed in only one of the units 0, 1, $+$, $-$ for each step of the (synchronous) simulation. In Section 5, we will show how the construction of Section 3 can be transformed into one which

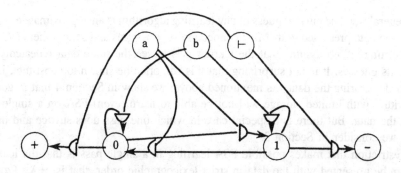

Figure 2. A connectionist parity network.

has a connectionist system learn to produce subnets like that of Figure 2. There have been some attempts (Williams, 1987) to extend conventional connectionist learning techniques to sequences, but our approach is quite different. It would be interesting to compare the various techniques.

More generally, we are interested in the range of applicability of various learning techniques and on how theoretical results can contribute to the development of learning machines. The starting point for the current investigation was the application of the theory of learning FSA to connectionist systems. As always, the assumptions in the two cases were quite different and had to be reconciled. There is, as yet, no precise specification on what constitutes a "connectionist" system, but there are a number of generally accepted criteria. The truism that any machine can be built from linear threshold elements is massively irrelevant. Connectionist architectures are characterized by highly parallel configurations of simple processors exchanging very simple messages. Any system having a small number of control streams, an interpreter or large amounts of passive storage is strongly anticonnectionist in spirit. It is this last characteristic that eliminated almost all the existing formal learning models as the basis for our study. Most work has assumed that the learning device can store all of the samples it has seen, and base its next guess on all this data. There have been a few studies on "iterative" learning where the guessing device can store only its last guess and the current sample (Wiehagen, 1976; Jantke & Beick, 1981; Osherson, et al., 1986). Some of the techniques from Jantke and Beick (1981) have been adapted to prove a negative result in Section 4. We show that a learning device using any finite amount of auxiliary memory cannot learn the Finite State Languages (FSL) from unordered presentations.

Another important requirement for a model to be connectionist is that it adapts. That is, a connectionist system should reflect its learning directly in the structure of the network. This is usually achieved by changing the weights on connections between processing elements. One also usually requires that the learning rule be local; a homunculus with a wire-wrap gun is decidedly unconnectionist. All of these criteria are based on abstractions of biological information processing and all were important in the development of this paper. The algorithm and proof of Section 3 do not mention them explicitly, but the results arose from these considerations. After a pedagogical transition in Section 5.1, Section 5.2 presents the outline of an FSL learner that is close to the connectionist spirit. Error tolerance, another connectionist canon, is only touched upon briefly but appears to present no fundamental difficulties.

In a general way, the current guess of any learning algorithm is an approximate encapsulation of the data presented to it. Most connectionist paradigms and some others (Valiant, 1984; Horning, 1969) assume that the learner gets to see the same data repeatedly and to refine its guesses. It is not surprising that this can often be shown to substitute, in the long run, for storing the data. As mentioned above, we show in Section 4 that in general an algorithm with limited storage will not be able to learn (even) FSA on a single pass through the data. But there is a special case in which one pass does suffice and that is the one we consider in Section 3.

The restriction that makes possible FSA learning in a single pass is that the learning algorithm be presented with the data in strict lexicographic order, that is, $\pm\lambda$, $\pm a$, $\pm b$, $\pm aa$, ... In this case the learner can construct an FSA, referred to also as the current guess, that exactly captures the sample seen so far. The FSA is nondeterministic, but consistent—every path through the FSA gives the same result for every sampled string considered so far. It turns out that this is a minimal state FSA consistent with the data and can thus be viewed as best guess to date. The idea of looking at strict lexicographic orders came to us in considering the algorithm of Rivest and Schapire (1987). Their procedure is equivalent to receiving $\pm$ samples in strict order. One would rarely expect such benign training in practical situations, but the general idea of starting with simpler examples is common.

Since the sample is presented in lexicographic order, our learning algorithm will be able to build up its guesses in a cumulative way. If the empty string is (is not) in the inferred language L, then the first guess is a machine with one accepting (rejecting) state. Each subsequent example is either consistent with the current guess, or leads to a new guess. The details of this comprise the learning algorithm of Section 3. When a new state is added to the current guess, a set of incoming and outgoing links to and from this new state are added. Consider the "even a's" language. With the sample $+\lambda$, the initial accepting state q_0 has links to itself under every letter. These links are all mutable and may later be deleted. When $-a$ is presented, the self-looping link under a is deleted and replaced by a permanent link to a new rejecting state q_1. We further add a mutable link from q_0 to q_1 under b, and the whole set of links from q_1. Figure 3 shows the guess for the "even a's" language after the initial sample $+\lambda$ and $-a$. The link from q_0 to q_1 under b is pruned when $+b$ is presented. $+aa$ will imply the deletion of the current self-link of q_1 under a, and $-ab$ will finally change the guess to that of Figure 1.

The remainder of the paper is divided into three major sections. Section 3 considers the general problem of learning FSA from lexicographically ordered strings. An algorithm

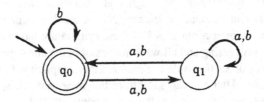

Figure 3. An intermediate guess for the parity FSA.

is presented and its space and time complexity are analyzed. The proof of correctness for this algorithm in Section 3.3 uses techniques from verification theory that have apparently not been used in the learning literature. In Section 4 we show that the strong assumption of lexicographic order is necessary—no machine with finite storage can learn the FSA from arbitrary samples. Section 5 undertakes the translation to the connectionist framework. This is done in two steps. First a distributed and modular, but still conventional version, is described. Then a transformation of this system to a connectionist network is outlined. Some general conclusions complete the paper.

2. Relation to previous work

The survey by Angluin and Smith (1983) is the best overall introduction to formal learning theory; we just note some of the most relevant work. Our learning algorithm (to be presented in the next section) identifies the minimum state deterministic FSA (DFSA) for any FSL *in the limit*: Eventually the guess will be the minimum state DFSA, but the learner has no way of knowing when this guess is found. The learner begins with no *a priori* knowledge. We can regard the sample data as coming from an unknown machine that identifies the inferred FSL. As stated above, our algorithm is an iterative strategy that uses at each stage a current encoding of the data considered so far, and the *single* current sample string. One of the crucial advantages of our algorithm is that the total amount of space used in the course of learning, for encoding any finite prefix of the sample, is polynomial in the size of the inferred minimumm-state DFSA. Any encoding of a target grammar requires $O(n^2)$ space, and our algorithm is $O(n^2)$.

Iterative learning strategies have been studied in Wiehagen (1976); Jantke and Beick (1981); and Osherson, et al. (1986). Jantke and Beick (1981) prove that there is a set of functions that can be identified in the limit by an iterative strategy, using the strict lexicographic presentations of the functions, but this set cannot be identified in the limit by an iterative strategy using arbitrary presentations. The proof can be slightly modified in order to prove that there is no iterative algorithm that can identify the FSL in the limit, using arbitrary representations for the languages. We generalize the definition of an iterative device to capture the ability to use any finite auxiliary memory in the course of learning. Hence, our result is stronger than that in Jantke and Beick (1981).

Gold (1967) gives algorithms for identifying FSA in the limit both for *resettable* and *nonresettable* machines where resettable machines are defined by the ability or need to "reset" the automation to some start state. These algorithms identify by means of enumeration. Each experiment is performed in succession, and in each stage all the experiments performed so far are used in order to construct the next guess. Consequently, the storage needed until the correct guess is reached is exponential in the size of the minimum state DFSA. The enumeration algorithm for resettable machines has the advantage (over our algorithm) that it does not specify the experiments to be performed; it can use any data that identifies the inferred FSL. This property is not preserved when the machines are nonresettable.

Gold (1972) introduces another learning technique for identifying a minimum state DFSA in the limit by experimenting with a resettable machine. This variation is called the *state characterization* method which is much simpler computationally. This technique specifies the experiments to be performed, and again has the disadvantage of having to monitor an infinitely increasing storage area.

Angluin (1987) bases her result upon the method of state characterization, and shows how to infer the minimum state DFSA by experimenting with the unknown automata (asking membership queries), and using an oracle that provides counterexamples to incorrect guesses. Using this additional information Angluin provides an algorithm that learns in time polynomial in the maximum length of any counterexample provided by the oracle, and the number of states in the minimum-state DFSA. Angluin's algorithm differs from our approach mainly in the use of an oracle that answers equivalence queries in addition to accepting or rejecting certain sample strings. The algorithm is comparable to ours in the sense that it uses experiments that are chosen at will.

Recently Rivest and Schapire (1987) presented a new approach to the problem of learning in the limit by experimenting with a nonresettable FSA. They introduce the notion of *diversity* which is the number of equivalence classes of *tests* (basically, an experiment from any possible state of the inferred machine). The learning algorithm uses a powerful oracle for determining the equivalence between tests, and finds the correct DFSA in time polynomial in the diversity. Since the lower bound on the diversity is log the number of states, and it is the best possible, this algorithm is practically interesting. Again, the experiments in this algorithm are chosen at will, and in fact they are a finite prefix of a lexicographically ordered sample of the inferred language.

Another variation of automation identification is that from a *given* finite subset of the input-output behavior. Bierman and Feldman (1972) discuss this approach. The learning strategy there includes an adjusted parameter for inferring DFSAs with varying degrees of accuracy, accomplished by algorithms with varying complexities. In general, Gold (1978) and Angluin (1978) prove that finding a DFSA of n states or less that is compatible with a given data is NP-complete. On the other hand, Trakhtenbrot and Barzdin (1973) and Angluin (1976) show that if the sample is *uniform-complete*, i.e., consists of all strings not exceeding a given length and no others, then there is a polynomial time algorithm (on the size of the whole sample) that finds the minimum state DFSA that is compatible with it. Note that the sample size is exponential in the size of the longest string in the sample. We can regard our algorithm as an alternative method for identifying the minmum state DFSA from a given uniform-complete sample. As stated above, our algorithm is much more efficient in space, since it does not access the whole sample, but rather refers to it in succession, and needs just a polynomial space in the number of states in the minimum state DFSA. The time needed for our algorithm is still polynomial in the size of the whole sample, though logarithmic in an amortized sense, as we show in Section 3.4.

Recently, Miri Sharon in her Master's thesis (Sharon, 1990) presented an algorithm for inferring FSA in the FS-IT model. The algorithm is based on Ibarra and Jiang, Lemma 1 (Ibarra & Jiang, 1988), and on the fact that the union, intersection, complement and reduction operations on finite automata, all require time and space polynomial in the size of the machine. Sharon's algorithm improves ours in that the time needed for each sample, as well as the storage, is polynomial in n, the size of the minimum state DFSA. The size of the inferred FSAs is not monotonically increasing; yet it is bounded by $6*n*n-3*n$. The design of the algorithm is inherently sequential and centralized.

3. Sequential version for learning FSA

3.1. Notation and definitions

We use the following notation and definitions:

- A *finite-state automata* (FSA) M is a 5-tuple $(Q, \Sigma, \delta, q_0, F)$ where
  - Q is a finite nonempty set of *states*.
  - Σ is a finite nonempty set of *letters*.
  - δ is a *transition function* that maps each pair (q, σ) to a set of states, where $q \in Q$ and $\sigma \in \Sigma$. This function can be represented by the set of *links* E so that $(p \, \sigma \, q) \in E$ iff $q \in \delta(p, \sigma)$. Each link is either *mutable* or *permanent*.

  δ can be naturally extended to any string $x \in \Sigma^*$ in the following way: $\delta(q, \lambda) = \{q\}$, and for every string $x \in \Sigma^*$ and for every letter $\sigma \in \Sigma$, $\delta(q, x\,\sigma) = \{p | (\exists \, r \in Q)(r \in \delta(q, x)$ and $p \in \delta(r, \sigma))\}$.
  - q_0 is the *initial* state, $q_0 \in Q$.
  - F is the set of *accepting* states, $F \subseteq Q$. (Q − F is called the set of rejecting states).
  - The *parity* $f(q)$ of a state $q \in Q$ is + if $q \in F$ and is − if $q \in Q − F$. By extension, assuming for some $q \in Q$ and $\sigma \in \Sigma$ that $\delta(q, \sigma) = \emptyset$, we define the parity of this state-symbol pair $f(q, \sigma)$ to be + if all successors of q under σ are + and − if they are all −. If all $r \in \delta(q, \sigma)$ do not have the same parity, then $f(q, \sigma)$ is undefined.
- A *deterministic* FSA (DFSA) is an FSA where the transition function δ is from $Q \times \Sigma$ into Q.
- The *language* L(M) accepted by a DFSA, M, is the set $\{x \in \Sigma^* \mid \delta(q_0, x) \in F\}$.
- Given a regular language L, we denote by M_L the (up to isomorphism) minimum state DFSA s.t. $L(M_L) = L$. Q_L is the state set of M_L.

The late lower case letters v, w, x, y, z will range over strings. Given a current FSA, the new string to be considered is denoted by w (the wrecker that may break the machine). Lower case letters p, q, r, s, t range over names of states. Whenever the current w wrecks the current guess, a new state, denoted by s (supplemental state) is added. σ, ϕ, ψ will range over letters, and i, j, k, m, n over the natural numbers.

- $M^x = (Q^x, \Sigma, \delta^x, q_0, F^x)$ is the FSA, referred to also as the *guess*, after the finite prefix $\pm\lambda, \ldots, \pm x$ of the complete lexicographically ordered sequence. E^x is the corresponding set of links.
- For $x \in \Sigma^*$, succ(x) stands for the string following x in the lexicographic order.
- The incremental construction of M^x admits for every state q, a unique string *minword*(q) that leads from q_0 to q using permanent links only. The path for *minword*(q) is referred to as the *basic* path to state q. These basic paths, which cover all the permanent links, form a spanning tree on the set Q^x.
- The guessing procedure also establishes for any M^x and any string y, unless y = *minword*(q) for some q, the existence of a unique state p, a letter ϕ and a string $z \in \Sigma^*$, such that y = *minword*(p)ϕz, and all the links from p under ϕ are mutable links. We

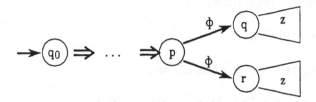

Figure 4. Tested state, tested letter and tested tail.

refer to these state, letter and string as the *tested state*, *tested letter* and *tested tail* (respectively) for y in M^x. Figure 4 shows the tree of all the paths for some string in some FSA, indicating the tested state p, tested letter ϕ and tested tail z.

We use the convention of representing a permanent link by $\Rightarrow$ and a mutable link by $\rightarrow$.

- For a given M^x and a word y, a path for y in M^x is *right* if it ends with an accepting state and $y \in L$, or it ends with a rejecting state and $y \notin L$. Otherwise, this path is called *wrong*.
- For two strings x, $y \in \Sigma^*$, and a language L, $x =_L y$ if both strings are in L, or both are not in L.

3.2. The learning algorithm

Let L be the regular language to be incrementally learned. Initially M^λ is constructed according to the first example $\pm\lambda$. $Q^\lambda = \{q_0\}$; $E^\lambda = \{q_0 \sigma q_0 \mid \sigma \in \Sigma\}$ and each link is mutable. If $\lambda \in L$, then q_0 is an accepting state, otherwise it is a rejecting one. *minword*(q_0) is set to λ.

Given M^x, the value *minword*(q) for every $q \in Q^x$, and a new string $\pm w$, $w = succ(x)$, the learning algorithm for constructing the new M^w is given in Figure 5. The algorithm is annotated with some important assertions (invariants in some control points) written between set brackets $\{\ldots\}$.

The subroutine *delete-bad-paths* (M, y, accept) is a procedure that constructs a new FSA out of the given M, in which every path for y leads to an accepting state iff accept = true. In the case y = w, *delete-bad-paths* breaks all wrong paths (if any) for w in M. In the case y < w, the paths in M are checked against the behavior of the old machine old-M. In any case, each bad path for y in M is broken by deleting its *first mutable link*. Note that all the first mutable links along bad paths are from the same tested state p for y in M. Furthermore, if all the paths for y in M are bad (and we will show that this can happen only if y = w), then after the execution of *delete-bad-paths* there will be no link from p under the tested letter for y in M. Such an execution will be followed by an execution of *insert-state*.

The procedure *insert-state* constructs a new FSA by extending the given spanning tree defined by the permanent links in old-M. A new state s is added. Let p and ϕ be the tested state and tested letter for w in old-M. Note again that all the mutable links from p under ϕ

```
begin
    old-M ← Mˣ;
    if w ∈ L then accept-w ← true else accept-w ← false;
    new-M ← delete-bad-paths (old-M, w, accept-w); /* Drop mutable links */
        {all the paths for w in new-M are right}
    if there is no path for w in new-M
    then {all the paths for w in Mˣ are wrong}
        {old-M is consistent with all strings up through x}
        repeat
            new-M ← insert-state; /* Insert new state, a new permanent link, and */
                                  /* new mutable links */
            {new-M may be inconsistent with previous strings}
            y ← λ;
            while succ(y) < w /* check against all previous strings */
                begin
                    y ← succ(y);
                    if all the paths for y in old-M lead to accepting states
                    then accept ← true
                    else accept ← false;
                    {accept is true if and only if y ∈ L}
                    {there exists a right path for y in new-M}
                    new-M ← delete-bad-paths (new-M, y, accept) /* Drop mutable links */
                    {new-M is now correct with respect to the strings λ, ..., y}
                end;
            old-M ← new-M;
            {old-M is consistent with all strings up through x}
            new-M ← delete-bad-paths (new-M, w, accept-w) /* Drop mutable links */
        until there exists a path for w in new-M;
        output new-M {Mʷ will be the new FSA new-M}
end
```

Figure 5. The learning algorithm for constructing M^w.

had been deleted in the last execution of *delete-bad-paths*. A new permanent link ($p \ \phi \ s$) is added. *minword*(s) is set to *minword*(p)ϕ. The parity of s is set under the following rule: If *minword*(s) = w, then s is an accepting state iff accept = true. In other words, in that case the parity of s is opposite to those states at the ends of the paths for w in old-M. If *minword*(s) < w then s is an accepting state iff all the paths for *minword*(s) in old-M end with accepting states. Next, mutable links to and from the new state s are added according to the following rule: For any existing state q, and for any letter σ, if *minword*(s)σ > *minword*(q), then add the mutable link (s σ q). Also, in the other direction, if the current links from q under σ are all mutable, and *minword*(q)σ > *minword*(s), add the mutable link (q σ s). Note that this rule adds (for q = s) all possible self links for the new state s. In other words, for every letter σ, the mutable link (s σ s) exists after *insert-state*.

Given M^x and w = succ(x), if all the paths for w in M^x are wrong, then the repeat loop takes place. This loop defines the extension process, which is a repetition of one or more applications of the *insert-state* procedure. It is easy to see that there will be at most $|w| - |minword(p)|$ insertions (application of *insert-state*), where p is the tested state for w in M^x. Suppose there are i insertions between M^x and M^w, each adding a new state. We can

13

refer to a sequence of length i of machines: M_0^w, M_1^w, ..., M_{i-1}^w, each of which is the old-M at the beginning of the repeat loop. M_0^w is the old-M as set to M^x at the beginning, the others are iteratively set in the body of the repeat loop. For every j, $0 \leq j \leq i - 1$, the execution of *insert-state* defines a new machine out of M_j^w, referred to as $M_j^w(\lambda)$. Thereafter, for every j, $0 \leq j \leq i - 1$, and for every y, $\text{succ}(\lambda) \leq y < w$, the execution of *delete-bad-paths* within the while loop defines a new machine (possibly the same as the preceding one), referred to as $M_j^w(y)$, indicating that this machine is ensured to be consistent with those strings up through y.

The algorithm was successfully implemented as a student course project by Lori Cohn, using C.

Before going on to the correctness proof ot his algorithm, we will discuss an example over the alphabet $\Sigma = \{a, b\}$. Suppose that the unknown language L is "number of a's is at most 1, and the number of b's is at least 1" or $bb^*(\lambda + ab^*) + abb^*$ as given by a regular expression. Figure 6 below shows some of the guesses.

Initially, since the first input example is $-\lambda$, q_0 is a rejecting state, having both (a, b) self-loop mutable links. For the next example $-a$, *delete-bad-paths* does not change the machine, hence M^a is the same as M^λ. When $+b$ is encountered, the mutable link $(q_0\, b\, q_0)$ is deleted, and the repeat loop takes place $(M_0^{\,b} = M^a)$. A new state $s = q_1$ is added, and a new permanent link $(q_0\, b\, q_1)$ is added. *minword*(q_1) is set to b. Since *minword*(q_1) is the current example, q_1 gets the opposite parity from that of q_0. Hence, q_1 is an accepting state. The new mutable links are $(q_1\, a\, q_1)$, $(q_1\, b\, q_1)$, $(q_1\, a\, q_0)$ and $(q_1\, b\, q_0)$. Note that $(q_0\, a\, q_1)$ is not added, since $a = minword(q_0)a$ is less than $b = minword(q_1)$. The new machine is $M_0^b(\lambda)$. Since all the paths (there is only one) for a in $M_0^b(\lambda)$ are right with respect to the old machine M_0^b, we get that $M_0^b(\lambda) = M_0^b(a)$. The only path for b is right, hence M^b is $M_0^b(a)$. The examples $-aa$ and $+ab$ do not change the current guess. When $+ba$ is encountered, there exists a right path $\langle q_0\, b\, q_1, q_1\, a\, q_1 \rangle$ and there exists a wrong path $\langle q_0\, b\, q_1, q_1\, a\, q_0 \rangle$. The first (and only) mutable link $(q_1\, a\, q_0)$ along the wrong path is deleted. A similar treatment is involved for the example $+bb$. Note that at this stage M^{bb} is a DFSA, but obviously $L(M^{bb}) \neq L$.

The next string $-aaa$ does not change the current guess, but $-aab$ causes a new application of *insert-state*. A new state q_2 is added, with *minword*(q_2) being a. The string aa is the first string that changes the machine while testing $M_0^b(b)$ against the old machine M_0^{aab}. The new mutable link $(q_2\, a\, q_1)$ is deleted. Other new mutable links are deleted while retesting ab, ba and bb. The execution of *delete-bad-paths* on $M_0^{aab}(aaa)$ $(= M_0^{aab}(bb))$ deletes the two mutable links $(q_2\, a\, q_2)$ and $(q_2\, a\, q_0)$, hence causing the nonexistence of a path for aab in the new machine. Thus, a new insertion is applied, replacing the two mutable links $(q_2\, a\, q_2)$ and $(q_2\, a\, q_0)$ by a new permanent link from q_2 to a new state q_3 under a. The retesting of the strings a, b, and aa against $M_1^{aab}(= M_0^{aab}(aaa))$ cause no change (no deletion) on $M_1^{aab}(\lambda)$. Some of the new mutable links are deleted while retesting ab, ba, bb and aaa. In $M_1^{aab}(aaa)$ there exist three right paths for aab, and one wrong path, causing the deletion of $(q_3\, b\, q_1)$, yielding M^{aab}. Given this guess, M^{aab}, the only path for aba is wrong. Note that this path has two mutable links and the first one is now replaced by a permanent link to the new accepting state q_4. The retesting deletes some of the new mutable links just recently added within the insertion of q_4.

When $-baa$ is checked, given M^{abb}, there are three right paths and two wrong paths. The tested state is q_1, the tested letter is a, and the tested tail is a. The first mutable link along the wrong paths is $(q_1\ a\ q_1)$. Hence, this link is deleted leaving $(q_1\ a\ q_4)$ as the only mutable link from q_1 under a. This link is the first mutable link along the three right paths. When $-aaab$ is checked, given M^{aaaa}, there are again three right paths and two wrong paths. This time, two mutable links $(q_3\ a\ q_0)$ and $(q_3\ a\ q_2)$ are deleted, each breaking a different wrong path. Note that this deletion leaves only one right path for $aaab$, as the two deleted links served as second mutable links along two of the original right paths for $aaab$ in M^{aaaa}. The correctness proff will show that after the deletion process, there exists at least one right path for the current sample. Finally, M^{abba} accepts the langauge L.

3.3. The correctness proof

Next we prove that the guessing procedure is correct. We first show that each M^x satisfies a set of invariants. Consider the following constraints for a FSA $M = (Q, \Sigma, \delta, q_0, F)$ and a given string $x \in \Sigma^*$:

(1) *Consistency*:
   $\forall y \leq x$, all the paths for y are right.
(2) *Completeness*:
   $(\forall\ q \in Q)(\forall\ \sigma \in \Sigma)$
   $((\exists\ r \in Q)(\delta(q, \sigma) = \{r\}$ and $(q\ \sigma\ r)$ is a permanent link)
   or
   $(\exists\ Q' \subseteq Q)(\delta(q, \sigma) = Q'$ and $Q' \neq \emptyset$ and $(\forall r \in Q')((q\ \sigma\ r)$ is a mutable link)))
(3) *Separability*:
   $(\forall q, r \in Q \mid q \neq r)(\exists\ y \in \Sigma^*)(minword(q)y \neq_L minword(r)y$ and $minword(q)y \leq x$ and $minword(r)y \leq x)$
(4) *Minimality*:
   $((\forall q \in Q)(\forall y \in \Sigma^*) \mid q \in \delta\ (q_0, y))(y \geq minword(q))$
(5) *Minword-property*:
   $(\forall q \in Q)(x \geq minword(q)$ and there is a unique string, namely $minword(q)$, that has a path from q_0 to q using permanent links only)

Note that some properties refer to a designated sample string x. We say that M y-satisfies a set of properties if whenever some property in this set relates to x, and y is substituted, then M satisfies the corresponding conjunct of properties.

The consistency constraint is the obvious invariant we would expect the learning algorithm to maintain. The completeness constraint means that for every state and for every letter there is either a permanent link that exits this state under this letter, or else there is a non-empty set of links leaving the state under this letter, all of which are mutable. The separability constraint together with Myhill-Nerode theorem (Hopcroft & Ullman, 1979) will lead to the claim that for each sample x, the number of states in M^x is at most the number of states in the minimum-state DFSA for the inferred language L. This can be established

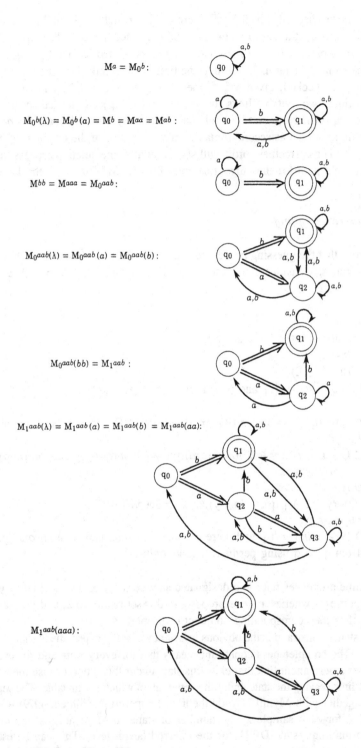

$M^a = M_0{}^b:$

$M_0{}^b(\lambda) = M_0{}^b(a) = M^b = M^{aa} = M^{ab}:$

$M^{bb} = M^{aaa} = M_0{}^{aab}:$

$M_0{}^{aab}(\lambda) = M_0{}^{aab}(a) = M_0{}^{aab}(b):$

$M_0{}^{aab}(bb) = M_1{}^{aab}:$

$M_1{}^{aab}(\lambda) = M_1{}^{aab}(a) = M_1{}^{aab}(b) = M_1{}^{aab}(aa):$

$M_1{}^{aab}(aaa):$

Figure 6. Learning bb* (λ + ab*) + abb*.

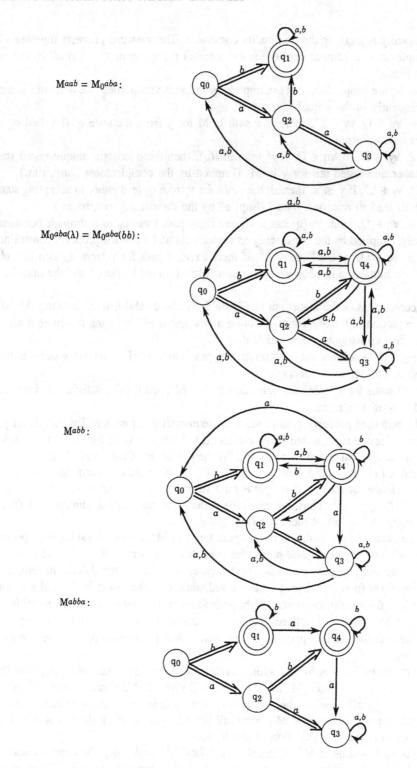

$M_{aab} = M_0aba:$

$M_0aba(\lambda) = M_0aba(bb):$

$M_{abb}:$

$M_{abba}:$

Figure 6. continued

by continually preserving the minimality constraint. The minword-property together with the completeness constraint implies the existence of the spanning tree formed by the permanent links.

Following are simple facts that are implied by the above properties. We will refer to these facts frequently in the sequel.

Fact 1. $\forall q \in Q$, $\forall y \in \Sigma^*$, there is a path in M for y from the state q. (Implied by the completeness constraint.)

Fact 2. $\forall y \in \Sigma^*$, if $\forall q \in Q$, $y \neq minword(q)$, then there exists a unique tested state, tested letter and tested tail for y in M. (Implied by the completeness constraint.)

Fact 3. $\forall y \in \Sigma^*$, if $y \leq x$, then all the paths for y from q_0 lead either to accepting states, or they all lead to rejecting states. (Implied by the consistency constraint.)

Fact 4. $\forall q \in Q$, $minword(q)$ has a unique right path from q_0 to q through permanent links only. (Implied by the consistency, completeness and minword-property constraints.)

Fact 5. $\forall q \in Q$, $\forall y \in \Sigma^*$, $\forall z \in \Sigma^*$, if there exists a path for y from q_0 to q that uses at least one mutable link, then $yz > minword(q)z$. (Implied by *Fact 4* and the minimality constraint.)

The correctness of the algorithm of Figure 5 will be established by showing that after each completion of the algorithm, yielding a new guess M^w by using the current sample $\pm w$, the five constraints are w-satisfied.

Clearly, M^λ λ-satisfies the constraints. Suppose (inductively) that M^x x-satisfies these constraints, and let $w = succ(x)$.

If all the paths for w in M^x are right, then $M^w = M^x$, and if M^x x-satisfies the invariants, then M^w w-satisfies them.

By the minword-property constraint, $w > minword(q)$ for each q. By *Fact 2*, let p, ϕ and z be the tested state, tested letter and tested tail (respectively) for w in M^x. Consider any of the states r, such that $(p \, \phi \, r) \in E^x$. By the definition of the tested elements, $(p \, \phi \, r)$ is a mutable link. By *Fact 5*, $minword(r)z \leq w$. Therefore, $minword(r)z$ has already been checked. Hence, by *Fact 3*, all the paths for z from r behave the same, i.e., $\delta^x(r, z) \subseteq F^x$ or $\delta^x(r, z) \subseteq Q^x - F^x$. Thus, either all the paths for w that use the mutable link $(p \, \phi \, r)$ are wrong paths, or all of them are right paths.

If there exist a wrong path and a right path for w in M^x, then by breaking each possible wrong path for w by *delete-bad-paths*, the consistency constraint is w-satisfied in M^w. To establish the completeness constraint in this case, note that all the deleted mutable links are of the form $(p \, \phi \, r)$, where p is the tested state, ϕ is the tested letter, and r is some state in M^x. Because there exists a right path for w in M^x, there must be a mutable link $(p \, \phi \, r')$ in M^x that is not deleted, and so M^w satisfies the completeness constraint. The other three constraints, separability, minimality and minword-property are obviously w-satisfied.

If all the paths for w in M^x are wrong, the expansion process takes place. Suppose there are i insertions in between M^x and M^w. We will show that the intermediate FSAs M_0^w, M_1^w, ..., M_{i-1}^w all x-satisfy the consistency, the completeness, the minimality and the minword-property constraints. Moreover, all the paths for w in M_j^w ($0 \leq j \leq i - 1$) are wrong, causing the re-application of *insert-state*.

M_0^w, being the same as M^x, obviously x-satisfies the consistency, the completeness, the minimality and the minword-property constraints, and all the paths for w in M_0^w are wrong.

Suppose for the moment that M_j^w, $0 \le j \le i - 1$, x-satisfies these four constraints. By the minword-property constraint, $w > minword(q)$ for each q. By *Fact 2*, let p, ϕ and z be the tested state, tested letter and tested tail (respectively) for w in M_j^w. Let s be the new state in $M_j^w(\lambda)$. In constructing $M_j^w(\lambda)$, the whole set of mutable links from p under ϕ in M_j^w are deleted but they are replaced by the new permanent link, (p ϕ s). This, plus the fact that we add all possible self-looping links for the new state s, establishes the part of the completeness constraint that ensures a nonempty set of links for each state and each letter. The other part—indicating that this set is either a singleton of a permanent link, or a set of mutable links—is easily implied by the construction. By the definition of *insert-state*, and the fact that permanent links are never deleted in *delete-bad-paths*, $M_j^w(\lambda)$ obviously w-satisfies the minword-property. As for the minimality constraint, suppose by way of contradiction that there exists a state q, such that the minimal string y that leads from q_0 to q is smaller than $minword(q)$. By the minword-property, this path uses at least one mutable link. By the minimality constraint of M_j^w, at least one of those mutable links is a new one just added while constructing $M_j^w(\lambda)$. Consider one of them, (r σ t). (Note that it is either the case that r = s or t = s.) From the way new links are added we immediately get a contradiction to the minimality assumption of y. Hence we conclude that $M_j^w(\lambda)$ w-satisfies the completeness, the minword-property and the minimality constraints.

The retesting process (within the while loop) checks the current machine $M_j^w(y)$ against the old machine M_j^w, that is assumed to be consistent up through x. The whole retesting process involves defining a sequence of FSAs: $M_j^w(\lambda)$, $M_j^w(succ(\lambda))$, $M_j^w(succ(succ(\lambda)))$, ..., $M_j^w(x)$. We will show that each $M_j^w(y)$ is consistent up through y, and that it w-satisfies the completeness, the minword-property and the minimality constraints. For this we need to refer to another inductive hypothesis, that will indicate, for each $y > \lambda$, the similarity between $M_j^w(y)$ and $M_j^w(\lambda)$. Let E' be the set of mutable links that were added while constructing $M_j^w(\lambda)$ from M_j^w. We will claim that each execution of *delete-bad-paths* along the construction of $M_j^w(y)$ out of $M_j^w(\lambda)$ deletes (if at all) only links from E'. Moreover, in the next paragraph we define a subset of E' that will definitely remain in $M_j^w(x)$. A link in this subset will be called a *necessary* link. Intuitively, these links will establish the existence of right paths on $M_j^w(\lambda)$ that reconstruct paths in M_j^w that use one of the mutable links in M_j^w from p under ϕ.

A link (s σ t) in $M_j^w(\lambda)$ is *necessary* iff t $\ne$ s, $w > minword(s)\sigma$, and there is a path for $minword(s)\sigma$ from q_0 to t in M_j^w. Figure 7 below shows how M_j^w and $M_j^w(\lambda)$ relate to each other with respect to p, s and t. The mutable link (s σ t) will definitely be added while constructing $M_j^w(\lambda)$, because by *Fact 5* applied on M_j^w, $minword(s)\sigma = minword(p)\phi\sigma > minword(t)$.

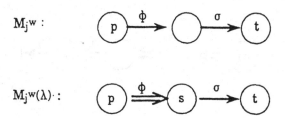

Figure 7. A necessary link (s σ t).

In order to prove that M^w_{j+1} x-satisfies the consistency, the completeness, the minword-property and the minimality constraints (given that M^w_j satisfies these conditions), we refer to another property, namely the *similarity constraint*: For each j, $0 \leq j \leq i - 1$, and each y, $\lambda \leq y \leq x$, $M^w_j(y)$ is the same as $M^w_j(\lambda)$ except for the removal of some of the new mutable links that were added while constructing $M^w_j(\lambda)$ out of M^w_j. Moreover, all the necessary links still exist in $M^w_j(y)$.

For every j, $0 \leq j \leq i - 1$, and for every y, $0 \leq y \leq x$, we will prove the following *intermediate invariant*: $M^w_j(y)$ satisfies the completeness, the minimality and the similarity constraints, it y-satisfies the consistency constraint, and w-satisfies the minword-property constraint.

We have already shown that $M^w_j(\lambda)$ preserves the completeness and the minimality constraints, and that it w-satisfies the minword-property constraint. Obviously, it satisfies the consistency up through λ, and the similarity (to itself). Thus, assuming inductively that $M^w_j(y)$, $\lambda \leq y < x$, satisfies the intermediate invariant, we need to show that so does $M^w_j(succ(y))$.

Recall that in the current execution of *delete-bad-paths*, the paths for succ(y) are checked against the behavior of succ(y) in M^w_j. If the current execution of *delete-bad-paths* causes no deletions (all the paths for succ(y) are right) then $M^w_j(succ(y))$ trivially maintains the intermediate invariant.

Otherwise, we show that it cannot be the case that all the paths for succ(y) in $M^w_j(y)$ are wrong. Moreover, if there exists a wrong path for succ(y) it will be broken by deleting one of the non-necessary new mutable links.

Assuming succ(y) has a wrong path in $M^w_j(y)$, we get by *Fact 4* that succ(y) $\neq$ *minword*(q) for each state q in $M^w_j(y)$. By *Fact 2*, let r, ψ and v be the tested state, the tested letter and the tested tail for succ(y) on $M^w_j(y)$. (As before, r is the last state reached by permanent links.) We distinguish between two possible cases:

1) r = s, i.e., succ(y) = *minword*(s)ψv.

Therefore, in M^w_j, the tested state and the tested letter for succ(y) were p and ϕ. Figure 8 shows the relation between M^w_j and $M^w_j(y)$ with respect to p and s. Let q, t be states such that (p ϕ q) and (q ψ t) are links in M^w_j. By the similarity constraint of $M^w_j(y)$, there must be some paths for succ(y) on $M^w_j(y)$ that use the existing necessary

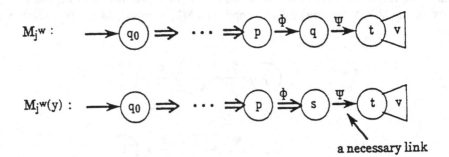

Figure 8. Proving the intermediate invariant: Case 1.

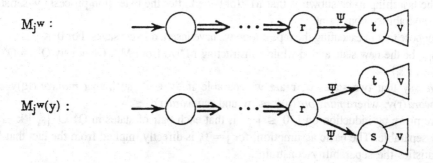

Figure 9. Proving the intermediate invariant: Case 2.

link (s ψ t). By *Fact 5* applied to $M_j^w(y)$, *minword*(t)v $<$ *minword*(s)ψv $=$ succ(y). Finally, by the consistency constraint of $M_j^w(y)$, all the paths for *minword*(t)v in $M_j^w(y)$ are right. Clearly, by *Fact 3* applied to M_j^w, *minword*(t)v $=_L$ succ(y), which implies in turn that all these paths for succ(y) in $M_j^w(y)$ that use the existing necessary link (s ψ t) must be right. Hence, this necessary link (s ψ t) will not be deleted. Other non-necessary mutable links from s that establish wrong paths for succ(y) (and there exists such a wrong one) will be deleted by the current execution of *delete-bad-paths*.

2) r $\neq$ s. Hence, r, ψ and v serve as the tested state, tested letter and tested tail for succ(y) on M_j^w also, and succ(y) $=$ *minword*(r)ψv. Figure 9 below indicates the relation between M_j^w and $M_j^w(y)$ in this case. Let t be a state in M_j^w such that some paths for succ(y) in M_j^w use the mutable link (r ψ t) (right after the permanent prefix). By the similarity constraint of $M_j^w(y)$, some paths for succ(y) on $M_j^w(y)$ use this existing mutable link. Again, by the consistency constraint, *Fact 3* and *Fact 5* applied to $M_j^w(y)$, all the paths in $M_j^w(y)$ for *minword*(t)v behave the same and are right. By *Fact 3* applied to M_j^w, *minword*(t)v $=_L$ succ(y). Hence, all the paths for succ(y) in $M_j^w(y)$ that use (r ψ t) are right, implying in turn that this (old) mutable link will not be deleted. By the assumption, there exists a wrong path for succ(y) in $M_j^w(y)$. Hence, there exists a new mutable link in $M_j^w(y)$, (r ψ s), that will be now deleted in order to break a bad path. This new mutable link is clearly a non-necessary one.

This terminates the discussion on the relation between $M_j^w(y)$ and $M_j^w($succ(y)$)$, and based on this we can easily conclude that the intermediate invariant is satisfied by $M_j^w($succ(y)$)$. Consequently, $M_j^w(x)$ satisfies this intermediate invariant, and in particular it is consistent up through x. For $0 \le j < i - 1$, $M_{j+1}^w = M_j^w(x)$. We get the desired hypothesis that this M_{j+1}^w x-satisfies the consistency, the completeness and the minimality constraints, and that all the paths for w in this new machine are wrong. Since $M_j^w(x)$ w-satisfies the minword-property constraint, and there exists a wrong path for w in $M_j^w(x)$, we get by *Fact 4* applied to $M_j^w(x)$ ($=M_{j+1}^w$) that M_{j+1}^w x-satisfies the minword-property constraint. For $j = i - 1$, we break all the wrong paths for w on $M_{i-1}^w(x)$ by deleting the first mutable links along them. By similar arguments as above, we get that the new M^w w-satisfies the consistency, the completeness, the minword-property and the minimality constraints.

The last thing to be shown is that M^w (obtained after the extension process) w-satisfies the separability constraint.

Suppose that in executing the repeat loop we have inserted i new states. For $0 \leq j \leq i - 1$, let s_{j+1} be the new state added while constructing $M_j^w(\lambda)$ from M_j^w. Obviously $Q^w = Q^x \cup \{s_1, \ldots s_i\}$.

We say that two states q, r are w-separable if $\exists y \in \Sigma^*$ such that $minword(q)y \neq_L minword(r)y$, where $minword(q)y \leq w$ and $minword(r)y \leq w$.

We prove by induction on j, $0 \leq j \leq i$, that each pair of states in $Q^x \cup \{s_k \mid k \leq j\}$ is w-separable. The basic assumption, for $j = 0$, is directly implied from the fact that M^x w-satisfies the separability constraint.

Assuming that each pair of states in $Q^x \cup \{s_k \mid k \leq j$ and $j < i\}$ is w-separable, we have to show that each state in this set is w-separable from s_{j+1}. Formally, let Q be the state set of M_j^w, we need to show that $(\forall q \in Q) (\exists y \in \Sigma^*) (minword(q)y \neq_L minword(s_{j+1})y$ and $minword(q)y \leq w$ and $minword(s_{j+1})y \leq w)$.

Let p, ϕ and z be the tested state, tested letter and tested tail for w in M_j^w, and $w = minword(s_{j+1})z$. Let q be an arbitrary state of M_j^w. We distinguish between the case where q was connected to p in M_j^w through the mutable link (p ϕ q), versus the case where they were not connected like this.

1) M_j^w has the mutable link (p ϕ q). By the corresponding execution of *delete-bad-paths*, this link will be deleted and replaced in $M_j^w(\lambda)$ by the new permanent link (p ϕ s_{j+1}). (p ϕ q) was deleted since all the paths for $w = minword(s_{j+1})z$ on M_j^w were wrong. By *Fact 5* applied on M_j^w, $minword(q)z < minword(p)\phi z = w$. By the consistency constraint of M_j^w, all the paths for $minword(q)z$ in M_j^w are right. Hence, $minword(s_{j+1})z \neq_L minword(q)z$, $minword(s_{j+1})z = w$ and $minword(q)z < w$.

2) The link (p ϕ q) does not exist in M_j^w. There can be two different sub-cases under this condition. The first one is that (p ϕ q) had never been added while constructing one of the previous FSAs. The second sub-case is that (p ϕ q) has once been deleted. Note that in each of the previous FSAs, the links that leave p under ϕ are always mutable links, and $M_j^w(\lambda)$ will be the first FSA having a permanent link from p under ϕ. Hence, if (p ϕ q) has been once added, and thereafter deleted, the deletion was due to an execution of *delete-bad-paths*.

   2.1) (p ϕ q) had never been added. Since it is not the case that there exists a permanent link from p under ϕ (establishing a possible reason for not adding (p ϕ q)), it must be that $minword(p)\phi < minword(q)$.

   Now clearly there exists a mutable link in M_j^w from p under ϕ. Let t be a state such that (p ϕ t) exists in M_j^w. By the induction hypothesis, q and t (being two distinct states in M_j^w) are w-separable, hence $\exists y \in \Sigma^*$, such that $minword(q)y \leq w$, $minword(t)y \leq w$ and $minword(q)y \neq_L minword(t)y$. Figure 10 shows the relation between M_j^w and $M_j^w(\lambda)$ with respect to p, q and t.

   By the initial assumption for this subcase, $minword(p)\phi < minword(q)$. Thus, since $minword(q)y \leq w$ and $minword(t)y \leq w$, we get that both $minword(t)y$ and $minword(p)\phi y$ are less than w. By the consistency constraint of M_j^w, all the paths for $minword(p)\phi y$, and all the paths for $minword(t)y$ are right, and clearly $minword(p)\phi y =_L minword(t)y$. Since y is separating between t and q, and

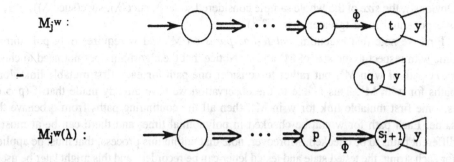

M_j^w :

$M_j^w(\lambda)$:

Figure 10. The case in which (p ϕ q) had never been added.

$minword(p)\phi = minword(s_{j+1})$, we can conclude that y is a sufficiently small tail separating between s_{j+1} and q.

2.2) The next subcase deals with the mutable link (p ϕ q) being deleted due to some string v, v ≤ x. (p ϕ q) has been the first mutable link along a wrong path for v on some previous FSA. Thus, there exists y $\in \Sigma^*$, such that v = $minword(p)\phi y$. The automata in which the decision to delete (p ϕ q) has been taken was obviously consistent with respect to $minword(q)y$, as $minword(q)y <$ v by the minimality constraint which is continually satisfied. This establishes the claim that $minword(p)\phi y \neq_L minword(q)y$. As $minword(q)y < minword(p)\phi$ y = $minword(s_{j+1})y ≤$ x, we get a perfect separating string for q and s_{j+1}.

This finishes the proof of the claim on the separability constraint.

Let R_L be the equivalence relation (Myhill-Nerode relation) associated with L, such that for x, y $\in \Sigma^*$, x R_L y iff ($\forall z \in \Sigma^*$)(xz =$_L$ yz). By Myhill-Nerode theorem, (Hopcroft & Ullman, 1979), $|Q_L|$ = number of equivalence classes of R_L.

By the separability constraint, $\forall x \in \Sigma^*$, $|Q^x| ≤ |Q_L|$. Thus, there exists $x^* \in \Sigma^*$, such that $\forall y ≥ x^*$ $|Q^y| = |Q^{x^*}|$ (after reaching M^{x^*} the extension process would never be applied again). Consider such an FSA, M^y where y ≥ x^*. Suppose there is a state q $\in Q^y$ for which there are at least two distinct mutable links (q σ r) and (q σ t). By the separability constraint, $\exists z \in \Sigma^*$ such that $minword(r)z \neq_L minword(t)z$. If both links still exist while considering the string $minword(q)\sigma z$ (by the consistency constraint, $minword(q)\sigma z >$ y) then, at this stage, one of them will definitely be deleted. Hence, eventually we will get a DFSA. Moreover, if M^x is a DFSA where $|Q^x| < |Q^L|$, then $\exists y >$ x such that the path for y in M^x is wrong. Therefore, we finally conclude that eventually we will get a minimum-state DFSA isomorphic to M_L. This completes the correctness proof.

Having proven the correctness of the learning algorithm, we now consider some properties of this process, mainly with respect to time and space complexity.

3.4. Complexity analysis

Given a current guess M^x, the value $minword(q)$ for every q $\in Q^x$ and the successor sample string ±w, we analyze first the time complexity. Let $|Q^x|$ = n, $|w|$ = m, and $|\Sigma|$ = a.

Obviously, the size of the whole sample considered so far λ, succ(λ), succ(succ(λ)), ..., w is exponential in m (greater than a^{m-1}).

First we note that executing *delete-bad-paths* on M^x and w requires only polynomial time with respect to the size of M^x and w. Notice that the algorithm does not need to check *every* path of w in M^x, but rather to consider one path for each first mutable link along paths for w in M^x. This is due to the observation we have already made that if (p ϕ q) is some first mutable link for w in M^x, then all the continuing paths from q behave the same. Each path for w can be checked in polynomial time, and there can be at most n different paths to be checked. Moreover, note that within this process, that must be applied for each string, the tested state and tested letter can be recorded, and this might later be used.

If *insert-state* is activated, then it can be done in polynomial time, gaining some efficiency by using the recorded tested state and tested letter. The dominating step is the addition of all possible new mutable links. This obviously involves considering such existing state q (and its *minword*), and each letter σ. Since, the length of w is m, *insert-state* can be repeatedly activated at most m times.

The retesting while loop is repeated for each string y, succ(λ) $\le$ y $\le$ x. For each such y, the condition that determines the value of the boolean variable accept can again be checked in polynomial time. Note that, due to the consistency constraint, only *one* path for y in the current old-M has to be checked to determine the behavior of the old machine with respect to y. The total retesting process (reaching the case for which succ(y) = w) can take exponential time in the size of the current example w, since all previous strings are checked, but it is still polynomial in the size of the whole sample considered so far. Such retesting happens very infrequently; in fact it is invoked once per state insertion. Therefore, taking into account the whole sample considered so far, the *amortized* cost of the retesting process is polynomial in the size of the current input. Finally, we conclude that the whole amount of time needed for the learning algorithm is polynomial in n in an amortized sense.

Furthermore, the extension process can be somewhat improved. First, in *insert-state*, we change the rule that adds a new mutable link from an old state q to the new state s under some letter σ. The new rule states that we add (q σ s) only when two conditions are met. The first condition is the original one, namely that the current links from q under σ are all mutable, and *minword*(q)σ > *minword*(s). The second condition is that either *minword*(q)σ $\ge$ w or *minword*(q)σ < w (so that the current parity of the pair (q, σ) is defined) and s is of the correct parity, i.e., f(s) = f(q, σ) in the current M_j^w. This rule is obviously correct, since if all the current links from q under σ are mutable, and *minword*(s) < *minword*(q)σ (so that (q σ s) would have been added under the previous rule), and moreover *minword*(q)σ < w and f(s) $\ne$ f(q, σ) (so that (q σ s) would not have been added under the new rule), then the retesting process will definitely prune this new mutable link while considering y = *minword*(q)σ. Hence, omitting those links immediately in *insert-state* might achieve some efficiency in the inspection of all possible paths for some y within the while loop.

A more significant improvement is due to the fact that within the retest process (the while loop) only new mutable links (to and from the new state) might be deleted as being first muable links along wrong paths for some y. Consequently, we need only check the following subset of the sample. For each state q and symbol σ, such that a new mutable link from q under σ has just been added within the last execution of *insert-state*, test the strings *minword*(q)σz that are smaller than w. Note that the new rule for adding new mutable links has now a more considerable impact on the performance, by omitting a whole set of strings from being checked. There are still n * $a^{|z|}$ such strings, but $|z|$ will usually be small and there even might be states for which no string will need to be tested.

As stated in the introduction, one of the major goals that motivated this work was to design an algorithm for learning an FSA that will require only a modest amount of memory, much less than the sample size which is exponential in m. Clearly, the storage needed for the current guess is proportional to $n^2 * a$, and the storage needed for all the values *minword*(q), for each state q, is proportional to n. The easiest way to envision the learning algorithm is to imagine that it uses two separate representations—one for new-M, and the other for old-M. Taken literally, this would double the size of the storage needed for the current guess. A more efficient solution is to indicate some links on the current guess as old ones, and thus analyze both machines on "one" representation. Another improvement might be gained (with respect to the amount of storage needed for the algorithm) by modifying the insertion process so as to avoid the need for storing *minword*(q) for each q. These values might as well be computed for every new state by linearly traversing the prefix-coded tree of the permanent links.

Let $|Q_L| = n_L$. As shown above, the current size n of the machine is at most n_L. Any two distinct states in M_L are obviously separable. Moreover, it can be shown that the shortest string that distinguishes between such pairs of states is at most of length n_L. Since each *minword*(q) is at most of length n_L, we can conclude that the maximum string after which the current guess is isomorphic to M_L is of length linear in n_L. In order to prove this, consider a guess M^x for which $|Q^x| = n_L$ and $|x| > 2 * n_L + 1$, and suppose (by way of contradiction) that M^x is not isomorphic to M_L. In other words M^x is non-deterministic, hence there exists a state p, such that there are at least two distinct mutable links from p under some letter ϕ. Let t and r be those states having incoming mutable links from p under ϕ. As indicated above, $|minword(p)| \leq n_L$, and there is a string of length at most n_L that separates between t and r. Hence, M^x admits two different paths for a word of length at most $2 * n_L + 1$, one that leads to an accepting state, and the other to a rejecting one. This obviously contradicts the consistency assumption of M^x. In summary, our algorithm could get by with space proportional to $a * n_L^2$ to store its guess plus m for the current string. This corresponds to the abstract notion of iterative learning of Wiehagen (1976).

4. Iterative learning using finite working storage

In this section we will formally characterize some "practical" properties of the learning algorithm introduced in Section 3. Taking into account the limitation of space in all realistic computations, the most important property of our algorithm is that for every sample, given in a lexicographic order, the algorithm uses a finite amount of space. The restriction on the strict order of the sample may seem to be too severe. We will show in this section that this restriction is necessary for learning with finite memory.

Our first definition follows the approach of Wiehagen (1976); Jantke & Beick (1981); and Osherson (1986):

Definition: An algorithm IT (iteratively) identifies a set of languages S iff for every $L \in S$, given the complete lexicographic sample $\langle \pm\lambda, \pm\text{succ}(\lambda), \pm\text{succ}(\text{succ}(\lambda)) \ldots \rangle$, the algorithm defines a sequence of finite machines $\langle M_0, M_1, M_2, \ldots \rangle$, such that $\forall i > 1$ M_i is obtained from the pair $(M_{i-1}, \pm x_i)$, where x_i is the i-th string in the sample, and $\exists j$ such that $\forall k \geq j$ $M_k = M_j$ and $L(M_j)$ (the language accepted by the machine M_j) is L.

It is easy to see that the algorithm of Section 3 meets the requirements of the IT definition. In other words, we exhibit an algorithm that IT identifies the FSL. Moreover, it finds a minimum-state DFSA for the inferred language L.

We now define a weaker characterization of a learning algorithm. We allow finite working storage in addition to that required for defining the current guess and sample string. The ultimate goal will be to show that the restriction on the order of the sample is necessary even for this kind of algorithm.

Definition: An algorithm FS-IT (iterative algorithm that uses finite storage) identifies a set of languages S iff for every $L \in S$, given the complete lexicographic sample $\langle \pm\lambda, \pm\text{succ}(\lambda)$, $\pm\text{succ}(\text{succ}(\lambda)), \ldots \rangle$, there exists a *finite* set of states Q, such that the algorithm defines a sequence of configurations $\langle (M_0, q_0), (M_1, q_1), (M_2, q_2), \ldots \rangle$ that satisfies the following: $\forall i$, $q_i \in Q$, $\forall i > 1$, (M_i, q_i) is obtained from the triple $(M_{i-1}, q_{i-1}, \pm x_i)$, where x_i is the i-th string in the sample, and $\exists j$ such that $\forall k \geq j$ $M_k = M_j$ and $L(M_j) = L$. Such a j (in the course of learning) is referred to as a "semi-stabilization" point. Note that the states within the configurations after the semi-stabilization point can still change.

Obviously, if an algorithm IT identifies S, then it also FS-IT identifies this set.

A (non-repetitive) complete sample for a language L is a sequence of its strings $\langle \pm x_1, \pm x_2, \pm x_3, \ldots \rangle$ such that $\forall i$, $x_i \in L$, $\forall i \neq j$, $x_i = x_j$, and $\forall x \in L$ $\exists i$ such that $x = x_i$. The ability to learn languages by presenting an *arbitrary* complete sample, rather than the strict lexicographic one, obviously strengthens the characterization of the learner. We denote the above situations by IT^{arb} and $\text{FS-IT}^{\text{arb}}$ if we do not require the sample to be in lexicographic order.

For a complete sample $\langle \pm x_1, \pm x_2, \pm x_3, \ldots \rangle$, an algorithm that $\text{FS-IT}^{\text{arb}}$ identifies a set of language S uses a finite set of states Q, and defines a sequence $\langle (M_0, q_0), (M_1, q_1), \ldots \rangle$. M_i and $q_i \in Q$ are referred to as the *current guess* and the *current state* after the finite prefix $\langle \pm x_1, \ldots, \pm x_i \rangle$, $\forall i \geq 0$.

Theorem. There is no algorithm that $\text{FS-IT}^{\text{arb}}$ identifies the finite state languages.

Proof. Suppose, to the contrary, that an algorithm A $\text{FS-IT}^{\text{arb}}$ identifies the FSLs. We will look for some FSLs that will lead to a contradiction.

Let L_0 be Σ^*, for some alphabet Σ.

For the lexicographically ordered sample for L_0, A defines the sequence $\langle (M_0^0, q_0^0), (M_1^0, q_1^0), \ldots \rangle$. By the definition of $\text{FS-IT}^{\text{arb}}$ and in particular due to the finiteness of the working storage, there exists some semi-stabilization point, i, that corresponds to some word x, such that the following two conditions are satisfied:

1) The current guess M_m^0, $\forall m > i$, is the same as M_i^0 and characterizes L_0. Call this guess M_{L_0}.
2) There are infinitely many m's, $m \geq i$, such that $q_m^0 = q_i^0$ (i.e., the state q_i^0 occurs infinitely often).

Let $L_1 = \{w \in \Sigma^* \mid w \leq x\}$.

For the lexicographically ordered sample for L_1 $\langle +\lambda, \ldots, +x, -succ(x), -\ldots \rangle$, A defines the sequence $\langle (M_0^1, q_0^1), (M_1^1, q_1^1), \ldots, (M_i^1, q_i^1), (M_{i+1}^1, q_{i+1}^1), \ldots \rangle$. Obviously, $\forall m, 0 \leq m \leq i, M_m^0 = M_m^1$ and $q_m^0 = q_m^1$. In particular $M_i^1 = M_{L_0}$ and $q_i^1 = q_i^0$. By the infinitely repeating property of q_i^0, and by the finiteness condition on the set $\{q_m^1 \mid m \geq 0\}$, q_i^0 must coincide with some q_m^1 infinitely often. In other words, there exists a j that corresponds to some word z, $j > i$, such that the following three conditions are satisfied:

1) $q_j^0 = q_i^0$.
2) $\forall m > j$ $M_m^1 = M_j^1$ and characterizes L_1. Call this guess M_{L_1}.
3) There are infinitely many m's, $m > j$, such that $q_m^0 = q_i^0$ and $q_m^1 = q_j^1$ (the pair of states q_i^0, q_j^1 appears infinitely many times at the same points for the strict ordered samples for L_0 and L_1).

Pick a place k, $k > j$, such that $q_k^0 = q_i^0$ and $q_k^1 = q_j^1$. The existence of such a place is established by the properties of the chosen j. Let y be the string at place k in the lexicographically ordered sample of any language over Σ. Note that $y > z$.

Let $L_2 = L_1 \cup \{w \in \Sigma^* \mid z < w \leq y\}$.

For the following ordered sample $\langle +\lambda, \ldots, +x, +succ(z), \ldots, +y, -succ(x), \ldots, -z, -succ(y), -succ(succ(y)), \ldots \rangle$, A defines the sequence $\langle (M_0^2, q_0^2), (M_1^2, q_1^2), \ldots \rangle$. From the above we get (see Figure 11):

1) After the finite prefix $\langle +\lambda, \ldots, +x \rangle$ the current guess is M_{L_0} and the state is q_i^0.
2) By the definition of z and y, after the finite prefix $\langle +\lambda, \ldots, +x, +succ(z), \ldots +y \rangle$ the current guess is still M_{L_0}, and the state is again q_i^0.
3) By the definition of z with respect to its occurrence in the strict ordered sample for L_1, after the finite prefix $\langle +\lambda, \ldots, +x, +succ(z), \ldots, +y, -succ(x), \ldots, -z \rangle$ the current guess M_k^2 is M_{L_1}, and $q_k^2 = q_j^1$.
4) By the property of q_j^1, $\forall m > k$, $M_m^2 = M_{L_1}$, and $q_m^2 = q_m^1$.

Hence A cannot FS-IT^{arb} identify L_2.

Thus we have shown formally that the FSL can be IT-identified (from lexicographically ordered samples) but cannot be IT^{arb}-identified. The theorems provide end-case results, but there is a wide range of possible presentation disciplines between IT and IT^{arb}. Obviously enough, our algorithm will still identify the FSL from presntations in which redundant strings happen to be missing. That is, any w such that $w = succ(x)$ and $M^w = M^x$ could be missing from the sample without effect. As stated above, for any FSL for which $|Q_L| = n$, namely an n-FSL, all the strings longer than $2 * n + 1$ will be redundant. Moreover, most strings of length at most $2 * n + 1$ could be missing from the sample.

$L_0:$ $+\lambda, ..., +x,$ ..., $+z,$ $+succ(z), ..., +y,$ ...
   $M_{L_0}, q_i{}^0$ $M_{L_0}, q_i{}^0$ $M_{L_0}, q_i{}^0$

$L_1:$ $+\lambda, ..., +x,$ $-succ(x), ..., -z,$ $-succ(z), ...,$ $-y,$
   $M_{L_0}, q_i{}^0$ $M_{L_1}, q_j{}^1$ $M_{L_1}, q_j{}^1$
   $-succ(y),$ $-succ(succ(y)),$ ...
   $M_{L_1}, q_{k+1}{}^1$ $M_{L_1}, q_{k+2}{}^1$

$L_2:$ $+\lambda, ..., +x,$ $+succ(z), ... +y,$ $-succ(x), ..., -z,$
   $M_{L_0}, q_i{}^0$ $M_{L_0}, q_i{}^0$ $M_{L_1}, q_j{}^1$
   $-succ(y),$ $-succ(succ(y)),$ ...
   $M_{L_1}, q_{k+1}{}^1$ $M_{L_1}, q_{k+2}{}^1$

Figure 11. Proving that A cannot FS-IT^{arb} identify L_2.

This follows because there are only $a * n^2$ links to be added or deleted and more than a^{2*n+1} strings of length $\leq 2 * n + 1$. Therefore a teacher could, in principle, get by with a greatly reduced presentation if she knew what to present. The same reduction could also be used in the retest phase of our algorithm.

We conjectured that perhaps for an ordered presentation of some particular subset, followed by an arbitrary sequence of the remaining strings, our algorithm would be still applicable. Let S_L be the set $\{minword(q) \mid q \in Q_L\}$, and let y be the maximal string in this set. We easily found a counterexample that shows that the subset $\{x \in \Sigma^* \mid x \leq y\}$ does not suffice, and it is not even the case that Q^y—the set of states in the last guess M^y—is the same as Q_L. We then examined a larger set. Let $S_{L'}$ be the set $\{x \mid (\exists p, q \in Q_L) (x = minword(p)z$ and $x \neq_L minword(q)z$ and $(\forall z' \in \Sigma^*) (minword(p)z' \neq_L minword(q)z' \Rightarrow x \leq minword(p)z'))\}$. The intuition behind this set is to include all the least strings that can distinguish between two distinct states in M_L. First we observe that $S_L \subseteq S_{L'}$. Consider some state p in Q_L. There must be some q in Q_L the parity of which differs from that of p, hence $minword(p) \neq_L minword(q)$. Obviously, for every $z' \in \Sigma^*$, $minword(p) \leq minword(p)z'$. Hence, $minword(p) \in S_{L'}$. Now, let y' be the maximal word in $S_{L'}$. The following counterexample shows that even the subset $\{x \in \Sigma^* \mid x \leq y'\}$ does not suffice. Let L be the language (given as an example in Section 3): "the number of a's is at most 1, and number of b's is at least 1." $S_{L'} = \{\lambda, a, b, aa, ab, ba, aab, aba\}$. After the finite prefix $\langle \lambda, -a, +b, ..., -aba \rangle$, M^{aba} is as shown in Figure 12.

Obviously, $M^{aba} \neq M_L$. Moreover, let $-bbaba$ be the next sample string. The rule for breaking all wrong paths by deleting the *first* mutable links along them cannot be applied in this case. The first mutable link along the bad path $\langle q_0 b q_1, q_1 b q_1, q_1 a q_1, q_1 b q_1, q_1 a q_1 \rangle$ is the link $\langle q_1 b q_1 \rangle$ that should remain in M_L. In fact, every other possible rule

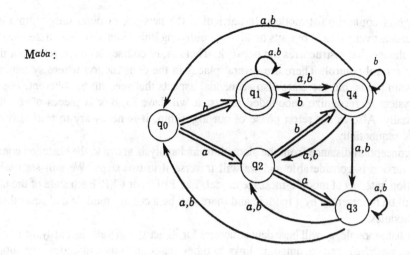

Figure 12. Applying the algorithm on an ordered subset.

that determines the mutable link to be deleted, according to its place in the bad path, would not work here. For every i = 1, 2, 3, 4, there is a bad path whose i-th mutable link exists in M_L.

It is an open question whether any characterization of the minimum training set exists.

It is also easy to see intuitively why finite storage learners will fail on arbitrary presentations. An arbitrary presentation can, for example, have only very long strings for a very long time and the learner has no idea what to make of them. This is the basic cause of the NP-completeness results of Gold (1978) and Angluin (1978) for minimal DFSA learning. On the other hand, if the learning device knew in advance the size, n, of the n-FSL, it might be able to collapse the long sample strings into equivalence classes. This is another open question. The realistic version of this is for the learner, which has finite storage, bounded by some polynomial in n, to limit its guesses to M with no more than n states.

There do seem to be some general consequences of the outcome of these open questions. If, as we surmise, knowing a bound on n for the target n-FSL does not permit FS-IT^{arb} with the finite storage being bounded by a polynomial in n, then learning simple examples first has inherent major advantages. If there are optimal training presentations, it will be interesting to understand their nature. As we will show in the next section, the algorithm of Section 3.1 works in a way that is compatible with connectionist and thus (at least for some people) with neural computation.

5. Distributed and connectionist versions

5.1. Distributed realization

As we discussed in the introduction, there is no generally accepted formalization of what precisely constitutes a connectionist model. In this section we show how the algorithm of Section 3.2 can be translated into a network of simple computing units that falls within the range of connectionist models. In particular, the network involves only simple units that broadcast very simple outputs on all their outgoing links. Learning is realized by local weight change that restructures the network. There is, of course, no interpreter, but there is some central control. There are several places in the construction where system-wide parallelism holds, but there are also sequential aspects that seem to be inherent. For any finite system to recognize unbounded inputs, it will have to look at pieces of the input sequentially. Also in the retest phase of our algorithm, it is necessary to test individual samples sequentially.

The conceptual distance from the algorithms and analysis above to the sketched connectionist version is considerable and we will traverse it in two steps. We will start with a realization in terms of module-message model, like PLITS or CSP. Each state of the target FSA will be represented by a module and there will be a control module and several other fixed modules.

Each state-module, q, will have data structures for its activation state, its parity, its minimal string $minword(q)$, and its outgoing links to other states and other modules. We suppose that the system is synchronous and that the control module broadcasts each letter of the input string, w, at the start of a major cycle. The first benefit of the parallel implementation is that all paths for the target string can be checked in parallel. Initially, q_0 is active. Each active module looks at the next symbol, σ, and sends an "activation" message along each of its outgoing links that correspond to σ. When these signals have been sent, any state that has not received new activation inactivates itself, and the states that have received such signals are active for the next cycle. There are three kinds of activation messages that are sent by states along paths of w. Recall the notation p and ϕ for the tested state and tested letter of w on the current machine. The first kind of an activation message is sent by those states along the basic path for $minword(p)$. This message indicates that no mutable link has yet occurred. Let p be the first state to use mutable outgoing link, p sends a message that encodes its identity, p, and that of the tested letter, ϕ. Each active module q that receives a message of the second kind, encoding the pair (p, ϕ), looks at the next symbol σ, and sends an activation message of the third kind along each of its outgoing links that correspond to σ. This message encodes the triple (p ϕ q), indicating a specific first mutable link along a path for w. All successor states that receive this kind of message send that same message. Note that each path is represented by its first mutable link, encoded within the message that passes through the corresponding suffix.

At the end of the string, marked by a terminator ⊢ or by a control signal, the states that are active report their corresponding first mutable links plus their parity $\pm$. This could be reported directly to the controller or (more connectionist) by sending activation to global variables (or modules) that represent good and bad strings. The control now compares the provided answer and if all reports are right, it goes on to the next string, as before.

If some parses are right and some are wrong, a deletion process corresponding to the subroutine *delete-bad-paths* must be executed. As mentioned and proved in Section 3, each of the first mutable links along paths for w corresponds either to a set of right paths, or to a set of wrong paths. The controller can obviously identify all paths by their first mutable links, and knows which ones were right or wrong from the reported parity. It then composes a message to all the "bad" mutable links, and sends this information. Each module that is the origin of a bad mutable link will then delete the corresponding outgoing link, thus breaking a bad path for w.

Finally, we need to model the state-addition and retest procedures. Clearly the control can easily discover if all paths are wrong. A new state is "recruited" by generating a new module, s, and initializing it with its parity, its path, *minword*(s) and with its permanent link. All this information can be easily generated using the data structure within the module that corresponds to the tested state p, namely the one that identifies itself as the source of all the bad mutable links to be deleted at that stage. With this information the new state/ module can establish links with old states following the strictures of *insert-state*. Again, this can be done in parallel except for the serialization within module s itself. The retest procedure is sequential, the controller cycles through the required strings and tests them against the old machine. The difference between the links of the old and new machines are also part of the data structures of the appropriate modules. Of course, within each string test, the parallel checking and deletion above still apply. Much of this will carry over to the connectionist version, but there are also several differences.

5.2. Connectionist realization

Connectionist models in the literature vary somewhat (Rumelhart & McClelland, 1986; Waltz & Feldman, 1986) but all are restricted to simple units that pass only numerical messages and always send the same number on each outgoing link. The links may have weights that modify the value being received and many models also allow conjunctive connections like we used in Figure 2. Rochester practice allows for a unit to have a small amount of internal data and to be in one of a small number of different "states" which we will denote here as "modes" to reduce confusion. The limited repertoire of connectionist systems forces the use of more elaborate structures than the previous version. We will present an outline of one such model.

The connectionist version of the FSA learner will have a control subnetwork that will sequence and modulate the basic learning net. There will be "registers," banks of units whose activation pattern represents a letter or a string of letters, like the top row of Figure 2. The basic process of testing a string against the current guess works as outlined in Figure 2. Each letter of the input string serves in turn as the gate on the conjunctive connections from state-unit to state-unit. The conjunction of activation of the prior state and the appropriate letter-unit leads to activation of the next state-units along appropriate links. One way to have the state-units turn off when they should is to have a just-sent mode. A unit in just-sent mode will inactivate itself (set its activation flag to zero) if it receives only a control signal for the next cycle.

A somewhat similar mechanism can be used to mark the first mutable link along each path. Suppose that permanent links have weight 1 and mutable links weight ½. Let the activation rule for a unit be as follows: The initial state q_0 always sends activation value 10 to start each string. If a subsequent state-unit sees an input value of 5, it knows that it is at the far end of the first mutable link in a path, and will record which input link was active. It will also send out a lower value, say 4. Units that receive either 2 or 4 will also send out 4 as a value. This effectively marks the receiving end of the first mutable link in every path, with a tagged input in the unit for that state marking the path.

Upon termination of the testing for some string, w, the global $+$ and $-$ units are compared by control with the answer provided (as activation of another Winner-Take-All pair). If all paths are right, then the next string is tried. If there are both right and wrong paths, the deletion process must occur. There is no obvious way to do this in parallel, but the following sequential scheme works. Assume that the mechanism includes a "buffer" that can record the input string w and another buffer that can be made to cycle through strings. Control recycles the input string (in delete mode) until the unit having the first mutable links is encountered. That is, a state-unit that is activated in delete mode and has its tag set sends a different signal which is detected by the control net. Then each such state is tested sequentially and the ones leading to wrong answers delete their corresponding incoming mutable link. Deletion can be just setting the weight to zero. There needs to be some mechanism for sequencing these states, e.g., enabling each state in sequential order.

The insertion process for the connectionist version involves even more technical details. It is reasonable to assume that the learning net has unused state-units that are connected to all the ones used thus far, one of these is "recruited" to be the new state s. It is not hard to determine which link to s should be permanent, it is the one from the state p with the first mutable links in the (wrong) parses of $w = minword(p)\phi z$. The string w could be reparsed and the output of 5 from the states that receive conjoined signals from p and ϕ could inform the new state, s, that it should set the weight of its active input link to be 1 (permanent). It is also reasonable to assume that the state-units can mark the current links from p under ϕ as "old," and thus to be used only in retest mode.

What we cannot assume is that state-units, q, can store the minimal string $minword(q)$ and compare it with $minword(s)$ to determine which $(q\ \sigma\ s)$ and $(s\ \sigma\ q)$ links should be added. Again, the apparent answer is to go sequential. We can assume that the control net has buffers for $minword(s)$ and $minword(q)$, $minword(s)$ is fixed for the addition process, but $minword(q)$ cycles through all the other existing states. The control net finds $minword(q)$ by testing strings, and with this in the buffer, the tests $minword(q)\sigma > minword(s)$ and $minword(s)\sigma > minword(q)$ can be carried out by the control net. The signal to break the appropriate links can be transmitted to the state-units involved. This leaves just the retesting process. The obvious way to handle this is to have the control net cycle through every string $y < w$ and test and correct the current guess. The basic process of testing a string and the deletion process work as before. Each unit needs to have an "old machine" and "new machine" mode and to know which links go with each. Each string less than w is tested in old machine mode and the answer is stored. Then the same string is tested in new machine mode and the deletion process is invoked for all wrong paths.

In this design, each unit would need internal data for recording its number, whether it is active, has an active first mutable link and is performing as the old or new machine. It would need "modes" for just-sent, for normal testing, for deletion, for recruiting and for pruning links. If we restrict ourselves to just state-less linear threshold elements, the complexity expands by an order of magnitude.

Yifat Weissberg's thesis project (Weissberg, 1990) describes an implementation of the learning algorithm in a connectionist network, using the Rochester Connectionist Simulator. Learning is achieved by weight changes. The rules by which the links change their weight are simple and local. Nevertheless, the whole implementation is quite complex, basically due to the net representation. The network consists of two parts, one represents an automaton, the other represents an example string. A special control unit connects these two parts. Every unit has a considerable number of sites (at which the incoming links arrive), so as to distinguish between different stages of the learning process. The basic algorithm has been slightly modified, to fit the connectionist style. The major changes are in the process of adding a new state to the current automaton. The thesis includes proofs for the correctness of the modified algorithm.

Since the details of each parse play an important part in the learning procedure, there are at least indirect connections with explanation-based learning. But the case of learning from a perfect, lexicographically ordered sample is a very special one. It is well worth exploring how the algorithms of this paper could be modified to deal with less controlled examples. An obvious approach is to change the delete process (of mutable links) to one that reduces weights to some non-zero value, perhaps halving them each time. The question of how to revise the rest of the network's operation to properly treat conflicting evidence is another topic worthy of further effort.

Acknowledgments

We thank John Case for providing useful comments on the manuscript.

This work was supported in part by ONR/DARPA Research contract No. N00014-82-K-0193 and in part by ONR Research contract no. N00014-84-K-0655.

References

Angluin, D. (1987). Learning regular sets from queries and counterexamples. *Information and Computation, 75,* 87–106.

Angluin, D. (1981). A note on the number of queries needed to identify regular languages. *Information and Control, 51,* 76–87.

Angluin, D. (1978). On the complexity of minimum inference of regular sets. *Information and Control, 39,* 337–350.

Angluin, D. (1976). *An application of the theory of computational complexity to the study of inductive inference.* Ph.D. dissertation, Department of Electrical Engineering & Computer Science, Univ. California, Berkeley.

Angluin, D., & Smith, C.H. (1983). Inductive inference: Theory and methods. *Computing Surveys, 15,* 237–269.

Biermann, A.W., & Feldman, J.A. (1978). On the synthesis of finite-state machines from samples of their behavior. *IEEE Trans. on Computers, C-21,* 592–597.

Brooks, R.A. (1987). Intelligence without representation. *Proceedings of the Conf. on Foundations of AI.* Cambridge, MA: MIT.

Feldman, J.A., & Ballard, D.H. (1982). Connectionist models and their properties. *Cognitive Science, 6,* 205–254.

Gold, E.M. (1978). Complexity of automaton identification from given data. *Information and Control, 37,* 302–320.

Gold, E.M. (1972). System identification via state characterization. *Automatica, 8,* 621–636.

Gold, E.M. (1967). Language identification in the limit. *Information and Control, 10,* 447–474.

Hinton, G.E. (1987). *Connectionist learning procedures* (TR CMU-CS-87-115). Pittsburgh, PA: Carnegie Mellon University, Computer Science Department.

Hopcroft, J.E., & Ullman, J.D. (1979). *Introduction to automata and formal languages.* Reading, MA: Addison-Wesley.

Horning, J.K. (1969). *A study of grammatical inference.* Ph.D. thesis, Stanford University.

Ibarra, O.H., & Jiang, T. (1988). Learning regular languages from counterexamples. *Proceedings of the 1988 Workshop on Computational Learning Theory* (pp. 371–385). Boston, MA.

Jantke, K.P., & Beick, H-R. (1981). Combining postulates of naturalness in inductive inference. *Journal of Information Processing and Cybernetics, 17,* 465–484.

Kearns, M., Li, J., Pitt, L., & Valiant, L. (1987). On the learnability of boolean formulae. *Proceedings of the 9th Annual ACM Symp. on Theory of Computing* (pp. 284–295). New York, NY.

Natarajan, B.K. (1987). On learning boolean functions. *Proceedings of the 9th Annual ACM Symp. on Theory of Computing* (pp. 296–394). New York, NY.

Osherson, D.N., Stob, M., & Weinstein, S. (1986). *Systems that learn: An introduction to learning theory for conginitive and computer scientists*. Cambridge, MA: MIT Press.

Rivest, R.L., & Schapire, R.E. (1987). A new approach to unsupervised learning in deterministic environments. *Proceedings of the 4th International Workshop on Machine Learning*. Irvine, CA.

Rivest, R.L., & Schapire, R.E. (1987). Diversity-based inference of finite automata. *Proceedings of the 28th Annual Symp. on Foundations of Computer Science*. Los Angeles, CA.

Rumelhart, D.E., & McClelland, J.L. (Eds.). (1986). *Parallel distributed processing, explorations in the microstructure of cognition*. Cambridge, MA: Bradford Books/MIT Press.

Sharon, M. (1990). *Learning automata*. M.Sc. Thesis in Computer Science, Technion, Haifa, Israel, (in Hebrew).

Trakhtenbrot, B.A., & Barzdin, Ya.M. (1973). *Finite automata*. Amsterdam: North-Holland.

Valiant, L.G. (1985). Learning disjunctions of conjunctions. *Proceedings of the 9th IJCAI* (pp. 560–566). Los Angeles, CA.

Valiant, L.G. (1984). A theory of the learnable. *CACM, 27*, 1134–1142.

Waltz, D., & Feldman, J.A. (Eds.). (1987). *Connectionist models and their implications*. Ablex Publishing Corp.

Weisberg, Y. (1990). *Iterative learning finite automata—Application by neural net*. M.Sc. thesis in Electrical Engineering, Technion, Haifa, Israel, (in Hebrew).

Wiehagen, R. (1976). Limeserkennung rekursiver funktionen durch spezielle strategien. *Elektronische Informationsverarbeitung und Kybernetik, 12*, 93–99.

Williams, R.J. (1987). *Reinforcement-learning connectionist systems* (TR NU-CCS-87-3).

Machine Learning, 7, 139–160 (1991)

SLUG: A Connectionist Architecture for Inferring the Structure of Finite-State Environments

MICHAEL C. MOZER   (MOZER@CS.COLORADO.EDU)
Department of Computer Science, and Institute of Congnitive Science, University of Colorado, Boulder, CO 80309-0430

JONATHAN BACHRACH   (BACHRACH@CS.UMASS.EDU)
Department of Computer and Information Science, University of Massachusetts, Amherst, MA 01003

Abstract. Consider a robot wandering around an unfamiliar environment, performing actions and sensing the resulting environmental states. The robot's task is to construct an internal model of its environment, a model that will allow it to predict the consequences of its actions and to determine what sequences of actions to take to reach particular goal states. Rivest and Schapire (1987a, 1987b; Schapire, 1988) have studied this problem and have designed a symbolic algorithm to strategically explore and infer the structure of "finite state" environments. The heart of this algorithm is a clever representation of the environment called an *update graph*. We have developed a connectionist implementation of the update graph using a highly-specialized network architecture. With back propagation learning and a trivial exploration strategy—choosing random actions—the network can outperform the Rivest and Schapire algorithm on simple problems. Perhaps the most interesting consequence of the connectionist approach is that, by relaxing the constraints imposed by a symbolic description, it suggests a more general representation of the update graph, thus allowing for greater flexibility in expressing potential solutions.

Keywords. Finite-state automata, automata induction, update graph, diversity-based inference, connectionism/neural networks, back propagation

Consider a robot placed in an unfamiliar environment. The robot is allowed to explore the environment by performing actions and sensing the resulting environmental state. The robot's task is to construct an internal model of the environment, a model that will allow it to predict the consequences of its actions and to determine what sequences of actions to take to reach particular goal states. This scenario is extremely general; it applies not only to physical environments, but also to abstract and artificial environments such as electronic devices (e.g., a VCR), computer programs (e.g., a text editor), and classical AI problem-solving domains (e.g., blocks world). Any agent—human or computer—that aims to manipulate its environment toward some desired end requires an internal representation of the environment. This is because, in any reasonably complex situation, the agent can directly perceive only a small fraction of the global environmental state at any time; the rest must be stored internally if the agent is to act effectively.

In this paper, we describe a connectionist network that learns the structure of its environment. The network architecture is based on a representation of finite-state automata developed by Rivest and Schapire (1987a, 1987b; Schapire, 1988). We begin by first describing several environments.

35

1. Sample environments

In each environment, the robot has a set of discrete *actions* it can execute to move from one environmental state to another. At each environmental state, a set of binary-valued *sensations* can be detected by the robot. Descriptions of five sample environments follow, the first three of which come from Rivest and Schapire.

1.1. The n-room world

The *n*-room world consists of *n* rooms arranged in a circular chain (Figure 1). Each room is connected to the two adjacent rooms. In each room is a light bulb and a light switch. The robot can sense whether the light in the room where it currently stands is on or off. The robot has three possible actions: move to the next room down the chain, move to the next room up the chain, and toggle the light switch in the current room.

1.2. The little prince world

The robot resides on the surface of a 2D planet. There are four distinct locations on the planet: north, south, east, and west. To the west, there is a rose; to the east, a volcano. The robot has two sensations, one indicating the presence of a rose at the current location, the other a volcano. The robot has available three actions: move to the next location in the direction it is currently facing, move to the next location away from the direction it is facing, and turn its head around to face in the opposite direction.

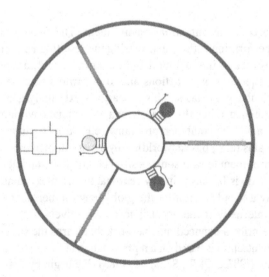

Figure 1. A three-room world.

1.3. The car radio world

The robot manipulates a car radio that can receive three stations, each of which plays a different type of music: top 40, classical, and jazz. The radio has two "preset" buttons labeled X and Y, as well as "forward seek" and "backward seek" buttons. There are three sensations, indicating the type of music played by the current station. The robot has six actions available: recall the station in preset X or Y, store the current station in preset X or Y, and search forward or backward to the next station from the current station.

1.4. The grid world

The grid world consists of an $n \times n$ grid of cells. Half of the cells possess distinct markings. The robot stands in one cell and can sense the marking in that cell, if any. There are thus $n^2/2$ sensations. The robot can take four actions: move to the next cell to the left, to the right, up, or down. Movement off one edge of the grid wraps around to the other side.

1.5. You figure it out

The above environments appear fairly simple partially because we have a wealth of world knowledge about light switches, radios, etc. For instance, we know that toggling a light switch undoes the effect of a previous toggle. The robot operates without the benefit of this background knowledge. To illustrate the abstract task that the robot faces, consider an environment with two actions, A and B, and one binary-valued sensation. Try to predict the sensations that will be obtained given the following sequence of actions.

Action: A A B B A B A B B B A B A A A B A B B A A B A B B

Resulting Sensation: 1 0 0 0 1 0 1 1 1 1 0 1 0 1 0 0 1 0 1 ? ? ? ? ? ?

If you give up, this is a simplified version of the n-room problem with only two rooms. Action A toggles the light switch, B moves from one room to the other, and the sensation indicates the state of the light in the current room. Initially, both lights are assumed to be off. People generally find this problem extremely challenging, if not insurmountably abstract, even when allowed to select actions and observe the consequences.

2. Modeling the environment with a finite-state automaton

The environments we wish to consider can be modeled by a finite-state automaton (*FSA*). Nodes of the FSA correspond to states of an environment. Labeled links connect each node of the automaton to other nodes and correspond to the actions that the robot can execute to move between environmental states. Associated with each node is a set of binary values: the sensations that can be detected by the robot in the corresponding environmental state.

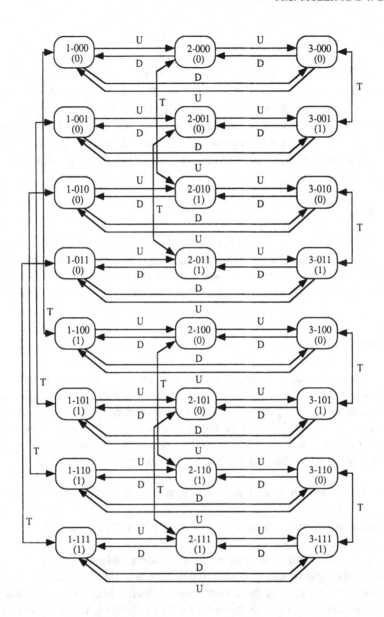

Figure 2. An FSA for the three-room world.

Figure 2 illustrates the FSA for a three-room world. Each node is labeled by the corresponding environmental state, coded in the form $r\text{-}s_1 s_2 s_3$, where r is the current room—1, 2, or 3—and s_i is the status of the light in room i—0 for off and 1 for on. The sensation associated with each node is written in parentheses. Links between nodes are labeled with one of the three actions: toggle (T), move up (U), or move down (D).

The FSA represents the underlying structure of the environment. If the FSA is known, one can predict the sensory consequences of any sequence of actions. Further, the FSA can be used to determine a sequence of actions required to obtain a certain goal state. For example, if the robot wishes to avoid light, it should follow a link or sequence of links for which the resulting sensation is 0.

Although one might try developing an algorithm to learn the FSA directly, there are several arguments against doing so (Schapire, 1988): (1) because many FSAs yield the same input/output behavior, there is no unique FSA; (2) knowing the structure of the FSA is unnecessary because we are only concerned with its input/output behavior; and, most importantly, (3) the FSA often does not capture redundancy inherent in the environment. As an example of this final point, in the n-room world, the T action has the same behavior independent of the current room number and the state of the lights in the other rooms, yet in the FSA of Figure 2, knowledge about "toggle" must be encoded for each room and in the context of the particular states of the other rooms. Thus, the simple semantics of an action like T are encoded repeatedly for each of the $n2^n$ distinct states.

3. Modeling the environment with an update graph

Rather than trying to learn the FSA, Rivest and Schapire suggest learning another representation of the environment called an *update graph*. The advantage of the update graph representation is that in environments with many regularities, the number of nodes in the update graph can be much smaller than in the FSA (e.g., $2n$ versus $n2^n$ for the n-room world).[1]

The Appendix summarizes Rivest and Schapire's formal definition of the update graph. Their definition is based on the notion of *tests* that can be performed on the environment, and the equivalence of different tests. In this section, we present an alternative, more intuitive view of the update graph that facilitates a connectionist interpretation of the graph.

Consider again the three-room world. To model this environment, the essential knowledge required is the status of the lights in the current room (CUR), the next room up from the current room (UP), and the next room down from the current room (DOWN). Assume the update graph has a node for each of these environmental variables. Further assume that each node has an associated value indicating whether the light in the particular room is on or off.

If we know the value of the variables in the current environmental state, what will their new values be after taking some action, say U? The new value of CUR becomes the previous value of UP; the new value of DOWN becomes the previous value of CUR; and in the three-room world, the new value of UP becomes the previous value of DOWN. As depicted in Figure 3a, this action thus results in shifting values around in the three nodes. This makes sense because moving up does not affect the status of any light, but it does alter the robot's position with respect to the three rooms. Figure 3b shows the analogous flow of information for the action D. Finally the action T should cause the status of the current room's light to be complemented while the other two rooms remain unaffected (Figure 3c). In Figure 3d, the three sets of links from Figure 3a–c have been superimposed and have been labeled with the associated action.

One final detail: The Rivest and Schapire update graph formalism does not make use of the "complementation" link. To avoid it, one may split each node into two values, one representing the status of a room and the other its complement (Figure 3e). Toggling thus involves exchanging the values of CUR and $\overline{\text{CUR}}$. Just as the values of CUR, UP, and DOWN must be shifted for the actions U and D, so must their complements.

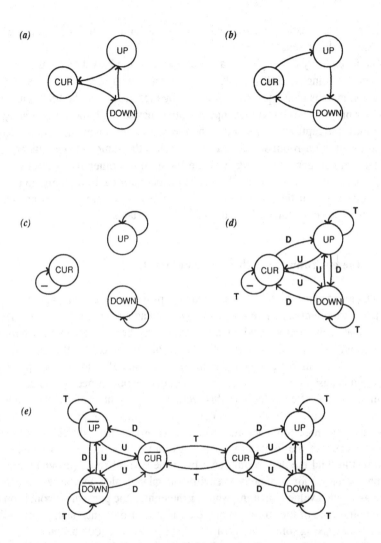

Figure 3. *(a)* Links between nodes indicating the desired information flow on performing the action U. CUR represents the status of the lights in the current room, UP the status of the lights in the next room up, and DOWN the status of the lights in the next room down. *(b)* Links between nodes indicating the desired information flow on performing the action D. *(c)* Links between nodes indicating the desired information on performing the action T. The "−" on the link from CUR to itself indicates that the value must be complemented. *(d)* Links from the three separate actions superimposed and labeled by the action. *(e)* The complementation link can be avoided by adding a set of nodes that represent the complements of the original set. This is the update graph for a three-room world.

Given the update graph in Figure 3e and the value of each node for the current environmental state, the result of any sequence of actions can be predicted simply by shifting values around in the graph. Thus, as far as predicting the input/output behavior of the environment is concerned, the update graph serves the same purpose as the FSA.

For every FSA, there exists a corresponding update graph. In fact, the update graph in Figure 3e might even be viewed as a distributed representation of the FSA in Figure 2.

In the FSA, each environmental state is represented by *one* "active" node. In the update graph, each environmental state is represented by a pattern of activity across the nodes.

One defining and nonobvious (from the current description) property of an update graph is that each node has exactly one incoming link for each action. For example, CUR gets input from $\overline{\text{CUR}}$ for the action T, from UP for U, and from DOWN for D. Table 1, which represents the update graph in a slightly different manner, provides another way of describing the unique-input property. The table shows node connectivity in the update graph, with a "1" indicating that two nodes are connected for a particular action, and "0" for no connection. The unique-input property specifies that there must be exactly one non-zero value in each row of each matrix.

Table 1. Alternative representation of update graph.

Update Graph Connectivity for *Move Up*

To Node	From Node					
	CUR	UP	DOWN	$\overline{\text{CUR}}$	$\overline{\text{UP}}$	$\overline{\text{DOWN}}$
CUR	0	1	0	0	0	0
UP	0	0	1	0	0	0
DOWN	1	0	0	0	0	0
$\overline{\text{CUR}}$	0	0	0	0	1	0
$\overline{\text{UP}}$	0	0	0	0	0	1
$\overline{\text{DOWN}}$	0	0	0	1	0	0

Update Graph Connectivity for *Move Down*

To Node	From Node					
	CUR	UP	DOWN	$\overline{\text{CUR}}$	$\overline{\text{UP}}$	$\overline{\text{DOWN}}$
CUR	0	0	1	0	0	0
UP	1	0	0	0	0	0
DOWN	0	1	0	0	0	0
$\overline{\text{CUR}}$	0	0	0	0	0	1
$\overline{\text{UP}}$	0	0	0	1	0	0
$\overline{\text{DOWN}}$	0	0	0	0	1	0

Update Graph Connectivity for *Toggle*

To Node	From Node					
	CUR	UP	DOWN	$\overline{\text{CUR}}$	$\overline{\text{UP}}$	$\overline{\text{DOWN}}$
CUR	0	0	0	1	0	0
UP	0	1	0	0	0	0
DOWN	0	0	1	0	0	0
$\overline{\text{CUR}}$	1	0	0	0	0	0
$\overline{\text{UP}}$	0	0	0	0	1	0
$\overline{\text{DOWN}}$	0	0	0	0	0	1

3.1. The Rivest and Schapire algorithm

Rivest and Schapire have developed a symbolic algorithm (hereafter, *the RS algorithm*) to strategically explore an environment and learn its update graph representation. They break the learning problem into two steps: (a) inferring the structure of the update graph, and (b) maneuvering the robot into an environmental state where the value of each node is known. Step (b) is relatively straightforward. Step (a) involves a method of experimentation to determine whether pairs of action sequences are equivalent in terms of their sensory outcomes. For a special class of environments, *permutation environments*, in which each action sequence has an inverse, the RS algorithm can infer the environmental structure— within an acceptable margin of error—by performing a number of actions polynomial in the number of update graph nodes and the number of alternative actions.

This polynomial bound is impressive, but in practice, Schapire (personal communication) achieves reasonable performance only by including heuristics that attempt to make better use of the information provided by the environment. To elaborate, the RS algorithm formulates explicit hypotheses about regularities in the environment and tests these hypotheses one or a relatively small number at a time. As a result, the algorithm may not make full use of the environmental feedback obtained. It thus seems worthwhile to consider alternative approaches that allow more efficient use of the environmental feedback, and hence, more efficient learning of the update graph. We have pursued a connectionist approach, which has shown quite promising results in preliminary experiments as well as a number of other powerful advantages. We detail these advantages below, but must first describe the basic approach.

4. Viewing the update graph as a connectionist network

SLUG is a connectionist network that performs subsymbolic learning of update graphs. Before the learning process itself can be described, however, we must first consider the desired outcome of learning. That is, what should SLUG look like following training if it is to behave as an update graph? Start by assuming one unit in SLUG for each node in the update graph. The activity level of the unit represents the boolean value associated with the update graph node. Some of these units serve as "outputs" of SLUG. For example, in the three-room world, the output of SLUG is the unit that represents the status of the current room. In other environments, there may be several sensations (e.g., the little prince world), in which case several output units are required.

What is the analog of the labeled links in the update graph? The labels indicate that values are to be sent down a link when a particular action occurs. In connectionist terms, the links should be *gated* by the action. To elaborate, we might include a set of units that represent the possible actions; these units act to multiplicatively gate the flow of activity between units in the update graph. Thus, when a particular action is to be performed, the corresponding action unit is activated, and the connections that are gated by this action become enabled.

If the action units form a local representation, i.e., only one is active at a time, exactly one set of connections is enabled at a time. Consequently, the gated connections can be replaced by a set of weight matrices, one per action (like those shown in Table 1). To predict the consequences of a particular action, say T, the weight matrix for T is simply plugged into the network and activity is allowed to propagate through the connections. Thus, SLUG is dynamically rewired contingent on the current action.

The effect of activity propagation should be that the new activity of a unit is the previous activity of some other unit. A linear activation function is sufficient to achieve this:

$$\mathbf{x}(t) = \mathbf{W}_{a(t)}\mathbf{x}(t - 1), \tag{1}$$

where $a(t)$ is the action selected at time t, $\mathbf{W}_{a(t)}$ is the weight matrix associated with this action, and $\mathbf{x}(t)$ is the activity vector that results from taking action $a(t)$. Assuming weight matrices like those shown in Table 1, which have zeroes in each row except for one connection of strength 1, the activiation rule will cause activity values to be copied around the network. Although linear activation functions are generally not appropriate for back propagation applications (Rumelhart, Hinton, & Williams, 1986), the architecture here permits such a simple function. SLUG is thus a linear system, which is extremely useful because it allows us to use the tools and methods of linear algebra for analyzing network behavior, as we show below.

5. Training SLUG

We have described how SLUG could be hand-wired to behave as an update graph, and now turn to the procedure used to *learn* the appropriate connection strengths. For expository purposes, assume that the number of units in the update graph is known in advance (this is not necessary, as we show below). A weight matrix is required for each action, with a potential non-zero connection between every pair of units. As in most connectionist learning procedures, the weight matrices are initialized to random values; the outcome of learning will be a set of matrices like those in Table 1.

If the network is to behave as an update graph, the critical constraint on the connectivity matrices is that each row of each weight matrix should have connection strengths of zero except for one value which is 1 (assuming Equation 1). To achieve this property, additional constraints are placed on the weights. We have explored a combination of three constraints:

(1) $\sum_j w_{aij}^2 = 1,$

(2) $\sum_j w_{aij} = 1,$ and

(3) $w_{aij} \geq 0,$

where w_{aij} is the connection strength to i from j for action a. If all three of these constraints are satisfied, the incoming weights to a unit are guaranteed to be all zeros except for one value which is 1. This can be intuited from Figure 4, which shows the constraints in a two-dimensional weight space. Constraint 1 requires that vectors lie on the unit circle,

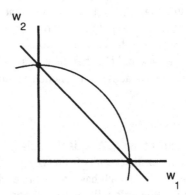

Figure 4. A two-dimensional space representing the weights, w_1 and w_2, feeding into a 2-input unit. The circle and line indicate the subregions specified by constraints 1 and 2, respectively. The two points of intersection are (1, 0) and (0, 1).

constraint 2 requires that vectors lie along the line $w_1 + w_2 = 1$, and constraint 3 requires that vectors lie in the first quadrant. Constraint 3 is redundant in a two-dimensional weight space but becomes necessary in higher dimensions.

Constraint 1 is satisfied by introducing a secondary error term,

$$E_{sec} = \sum_{a,i} (1 - \|\mathbf{w}_{ai}\|)^2,$$

where $\mathbf{w}_{ai}$ is the incoming weight vector to unit i for action a. The learning procedure attempts to minimize this error along with the primary error asociated with predicting environmental sensations. Constraints 2 and 3 are rigidly enforced by renormalizing the $\mathbf{w}_{ai}$ following each weight update. The normalization procedure finds the shortest distance projection from the updated weight vector to the hyperplane specified by constraint 2 that also satisfies constraint 3.

5.1. Details of the training procedure

Initially, a random weight matrix is generated for each action. Weights are selected from a uniform distribution in the range [0, 1] and are normalized to satisfy constraint 2. At each time step t, the following sequence of events transpires:

1. An action, $a(t)$, is selected at random.
2. The weight matrix for that action, $\mathbf{W}_{a(t)}$, is used to compute the activities at t, $\mathbf{x}(t)$, from the previous activities $\mathbf{x}(t - 1)$.
3. The selected action is performed on the environment and the resulting sensations are observed.
4. The observed sensations are compared with the sensations predicted by SLUG (i.e., the activities of units chosen to represent the sensations) to compute a measure of error. To this error is added the contribution of constraint 1.

5. The back propagation "unfolding-in-time" procedure (Rumelhart, Hinton, & Williams, 1986) is used to compute the derivative of the error with respect to weights at the current and earlier time steps, $\mathbf{W}_{a(t-i)}$, for $i = 0 \ldots \tau - 1$.
6. The weight matrices for each action are updated using the overall error gradient and then are renormalized to enforce constraints 2 and 3.
7. The temporal record of unit activities, $\mathbf{x}(t - i)$ for $i = 0 \ldots \tau$, which is maintained to permit back propagation in time, is updated to reflect the new weights. (See further explanation below.)
8. The activities of the output units at time t, which represent the predicted sensations, are replaced by the observed sensations.[2]

Steps 5–7 require further elaboration. The error measured at step t may be due to incorrect propagation of activities from step $t - 1$, which would call for modification of the weight matrix $\mathbf{W}_{a(t)}$. But the error may also be attributed to incorrect propagation of activities at earlier times. Thus, back propagation is used to assign blame to the weights at earlier times. One critical parameter of training is the amount of temporal history, τ, to consider. We have found that, for a particular problem, error propagation beyond a certain critical number of steps does not improve learning performance, although any fewer does indeed harm performance. In the results described below, we generally set τ for a particular problem to what appeared to be a safe limit: one less than the number of nodes in the update graph solution of the problem.

To back propagate error in time, a temporal record of unit activities is maintained. However, a problem arises with these activities following a weight update: the activities are no longer consistent with the weights—i.e., Equation 1 is violated. Because the error derivatives computed by back propagation are exact only when Equation 1 is satisfied, future weight updates based on the inconsistent activities are not assured of being correct. Empirically, we have found the algorithm extremely unstable if we do not address this problem.

In most situations where back propagation is applied to temporally-extended sequences, the sequences are of finite length. Consequently, it is possible to wait until the end of the sequence to update the weights, at which point consistency between activities and weights no longer matters because the system starts fresh at the beginning of the next sequence. In the present situation, however, the sequence of actions does not terminate. We thus were forced to consider alternative means of ensuring consistency. One approach we tried involved updating the weights only after every, say, 25 steps. Immediately following the update, the weights and activities are inconsistent, but after τ steps (when the inconsistent activities drop off the activity history record), consistency is once again achieved. A more successful approach involved updating the activities after each weight change to force consistency (step 7 of the list above). To do this, we propagated the earliest activities in the temporal record, $\mathbf{x}(t - \tau)$, forward again to time t, using the updated weight matrices.[3]

The issue of consistency arises because at no point in time is SLUG instructed as to the state of the environment. That is, instead of being given an activity vector as input, part of SLUG's learning task is to discover the appropriate activity vector. This might suggest a strategy of explicitly learning the activity vector, that is, performing gradient descent in both the weight space and activity space. However, our experiments indicate that this strategy does not improve SLUG's performance. One plausible explanation is the following.

If we perform gradient descent in weight space based on the error from a single trial, and then force activity-weight consistency, the updated output unit activities are guaranteed to be closer to the target values (assuming a sufficiently small learning rate and that the weight constraints have minor influence). Thus, the effect of this procedure is to reduce the error in the observable components of the activity vector, which is similar to performing gradient descent in activity space directly.

A final comment regarding the training procedure: In our simulations, learning performance was better with target activity levels of −1 and +1 (for *light is off* and *on*, respectively) rather than 0 and 1. One explanation for this is that random activations and random (nonnegative) connection strengths tend to cancel out in the −1/+1 case, but not in the 0/1 case.

6. Results

SLUG's architecture, dynamics, and training procedure are certainly nonstandard and nonintuitive. Our original experiments in this domain involved more standard recurrent connectionist architectures (e.g., Elman, 1988; Mozer, 1989) and were spectacularly unsuccessful. Many simple environments could not be learned, predictions were often inaccurate, and a great deal of training was required. Rivest and Schapire's update graph representation has thus proven beneficial in the development of SLUG. It provides strong constraints on network architecture, dynamics, and training procedure.

Figure 5 shows the weights in SLUG for the three-room world at three stages of learning. The "step" refers to how many actions the robot has taken, or equivalently, how many times the weights have been updated. The bottom diagram in the figure depicts a connectivity pattern identical to that presented in Table 1, and corresponds to the update graph of Figure 3e. To explain the correspondence, think of the diagram as being in the shape of a person who has a head, left and right arms, left and right legs, and a heart. For the action U, the head—the output unit—receives input from the left leg, the left leg from the heart, and the heart from the head, thereby forming a three-unit loop. The other three units— the left arm, right arm, and right leg—form a similar loop. For the action D, the same two loops are present but in the reverse direction. These two loops also appear in Figure 3e. For the action T, the left and right arms, heart, and left leg each keep their current value, while the head and the right leg exchange values. This corresponds to the exchange of values between the CUR and $\overline{\text{CUR}}$ nodes of the Figure 3e.

In addition to learning the update graph connectivity, SLUG has simultaneously learned the correct activity values associated with each node for the current state of the environment. Armed with this information, SLUG can predict the outcome of any sequence of actions. Indeed, the prediction error drops to zero, causing learning to cease and SLUG to become completely stable. Because the ultimate weights and activities are boolean, SLUG can predict infinitely far into the future with no degradation in performance (cf., Servan-Schreiber, Cleeremans, & McClelland, 1988).

Now for the bad news: SLUG does not converge for every set of random initial weights, and when it does, it requires on the order of 6,000 steps, much greater than the RS algorithm.[4] However, when the weight constraints are removed, SLUG converges without fail and in about 300 steps. It appears that only in extremely small environments do the weight

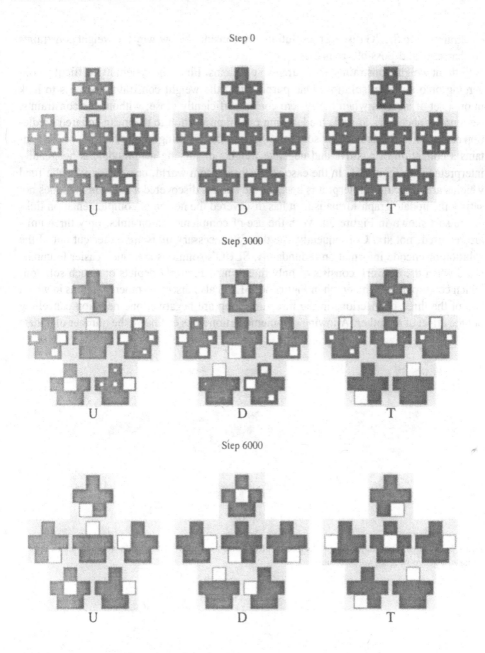

Figure 5. The three-room world. Weights learned by SLUG with six units at three stages of learning: step 0 reflects the initial random weights, step 3000 reflects the weights midway through learning, and step 6000 reflects the weights upon completion of learning. Each large diagram represents the weights corresponding to one of the three actions. Each small diagram contained within a large diagram represents the connection strengths feeding into a particular unit for a particular action. There are six units, hence six small diagrams. The output unit, which indicates the state of the light in the current room, is the protruding "head" of the large diagram. A white square in a particular position of a small diagram represents the strength of connection from the unit in the homologous position in the large diagram to the unit represented by the small diagram. The area of the square is proportional to the connection strength.

constraints help SLUG discover a solution. We consider below why the weight constraints are harmful and possible remedies.

Without weight constraints, there are two problems. First, the system has difficulty converging onto an exact solution. One purpose that the weight constraints serve is to lock in on a set of weights when the system comes sufficiently close; without the constraints, we found it necessary to scale the learning rate in proportion to the mean-squared prediction error to avoid overstepping solutions. Second the resulting weight matrix, which contains a collection of positive and negative weights of varying magnitudes, is not readily interpreted (see Figure 6). In the case of the three-room world, one reason why the final weights are difficult to interpret is because the net has discovered a solution that does not satisfy the update graph formalism; it has discovered the notion of complementation links of the sort shown in Figure 3d. With the use of complementation links, only three units are required, not six. Consequently, the three unnecessary units are either cut out of the solution or encode information redundantly. SLUG's solutions are much easier to understand when the network consists of only three units. Figure 7 depicts one such solution, which corresponds to the graph in Figure 3d. SLUG also discovers other solutions in which two of the three connections in the three-unit loop are negative, one negation cancelling out the effect of the other. Allowing complementation links can halve the number of update

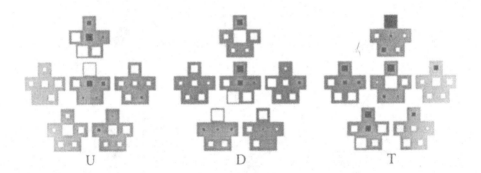

Figure 6. The three-room world. Weights learned by SLUG with six units without weight constraints. Black squares indicate negative weights, white positive.

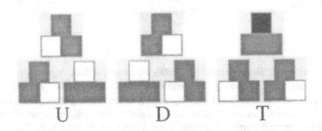

Figure 7. The three-room world. Weights learned by SLUG with three units without weight constraints. (The weights have been cleaned up slightly to make the result clearer.)

Table 2. Number of steps required to learn update graph.

Environment	RS Algorithm	SLUG
Little Prince World	200	91
Three-Room World	not available	298
Four-Room World	1,388	1,509
6×6 Grid World	not available	8,142
Car Radio World	27,695	8,167
32-Room World	52,436	fails

graph nodes required for many environments. This is one fairly direct extension of Rivest and Schapire's update graph formalism that SLUG suggests.

Table 2 compares the performance of the RS algorithm against that of SLUG without weight constraints for a sampling of environments.[5] Performance is measured in terms of the number of actions the robot must take before it is able to predict the outcome of subsequent actions, that is, the number of actions required to learn the update graph structure and the truth value associated with each node. The performance data reported for SLUG was the median over 25 replications of each simulation. SLUG was considered to have learned the task on a given trial if the correct predictions were made for at least the next 2,500 steps.

The learning rates used in our simulations were adjusted dynamically every 100 steps by averaging the current learning rate with a rate proportional to the mean squared error obtained on the last 100 steps. Several runs were made to determine what initial learning rate and constant of proportionality yielded the best performance. It turned out that performance was relatively invariant under a wide range of these parameters. Momentum did not appear to help.[6]

In simple environments, the connectionist update graph can outperform the RS algorithm. These results are quite surprising when considering that the action sequence used to train SLUG is generated at random, in contrast to the RS algorithm, which involves a strategy for exploring the environment. We conjecture that SLUG does as well as it does because it considers and updates many hypotheses in parallel at each time step. That is, after the outcome of a single action is observed, nearly all weights in SLUG are adjusted simultaneously.

In complex environments—ones in which the number of nodes in the update graph is quite large and the number of distinguishing environmental sensations is relatively small— SLUG does poorly. As an example of such, a 32-room world cannot be learned by SLUG whereas the RS algorithm succeeds. An intelligent exploration strategy seems necessary in complex environments: with a random exploration strategy, the time required to move from one state to a distant state becomes so large that links between the states cannot be established.

The 32-room world is extreme; all rooms are identical and the available sensory information is meager. Such an environment is quite unlike natural environments, which provide a relative abundance of sensory information to distinguish among environmental states. SLUG performs much better when more information about the environment can be sensed directly. For example, learning the 32-room world is trivial if SLUG is able to sense the states of all 32 rooms at once (the median number of steps to learn is only 1,209). The 6×6 grid world is another environment as large as the 32-room world in terms of the number of nodes SLUG requires, but it is much easier to learn because of the rich sensory information.

6.1. Noisy environments

The RS algorithm is not designed to handle environments with stochastic sensations. In contrast, SLUG's performance degrades gracefully in the presence of noise. For example, SLUG is able to learn the update graph for the little-prince world even when sensations are unreliable, say, when sensations are registered incorrectly 10% of the time. To train SLUG properly in noisy environments, however, the training procedure must be altered. If the observed sensation replaces SLUG's predicted sensation and the observed sensation is incorrectly registered, the values of nodes in the graph are disrupted and SLUG requires several noise-free steps to recover. Thus, a procedure might be used in which the predicted sensation is not completely replaced by the observed sensation, but rather some average of the two is computed; additionally, the average should be weighted towards the prediction as SLUG's performance improves.

6.2. Prior specification of update graph size

The RS algorithm requires an upper bound on the number of nodes in the update graph. The results presented in Table 2 are obtained when the RS algorithm knows exactly how many nodes are required in advance. The algorithm fails if it is given an upper bound less than the required number of nodes, and performance degrades as the upper bound increases above the required number. SLUG will also fail if it is given fewer units than are necessary for the task. However, performance does not appear to degrade as the number of units increases beyond the minimal number. Table 3 presents the median number of steps required

Table 3. Median number of steps to learn four-room world.

Units in SLUG	Number of Steps to Learn Update Graph
4	2,028
6	1,380
8	1,509
10	1,496
12	1,484
14	1,630
16	1,522
18	1,515
20	1,565

to learn the four-room world as the number of units in SLUG is varied. Although performance is independent of the number of units here, extraneous units greatly *improve* performance when the weight constraints are applied. Only 3 of 25 replications of the four-room world simulation with 8 units and weight constraints successfully learned the update graph (the simulation was terminated after 100,000 steps), whereas 21 of 25 replications succeeded when 16 units were used.

6.3. *Generalizing the update graph formalism*

Having described the overall performance of SLUG, we return to the issue of why weight constraints appear to harm performance. One straightforward explanation is that there are many possible solutions, only a small number of which correspond to update graphs. With weight constraints, SLUG is prevented from finding alternative solutions. One example of an alternative solution is the network with complementation links presented in Figure 7. Allowing complementation links can halve the number of update graph nodes required for many environments.

An even more radical generalization of the update graph formalism arises from the fact that SLUG is a linear system. Consider a set of weight matrices, $\{W_a\}$, that correspond to a particular update graph, i.e., each row of each matrix satisfies the constraint that all entries are zero except for one entry that is one (as in Table 1). Further, assume the vector x indicates the current values of each node. Given the previously-stated activation rule,

$$x(t) = W_{a(t)}x(t - 1),$$

one can construct an equivalent system,

$$x'(t) = W'_{a(t)}x'(t - 1)$$

by substituting

$$x'(t) \equiv Qx(t)$$

and

$$W'_{a(t)} \equiv QW_{a(t)}Q^*.$$

Here, the W_a have dimensions $n \times n$, the W'_a have dimensions $m \times m$, Q is any matrix of dimensions $m \times n$ and rank n, and Q^* is the left inverse of Q. The transformed system consisting of x' and the W'_a is isomorphic to the original system; one can determine a unique x for each x', and hence one can predict sensations in the same manner as before. However, the transformed system is different in one important respect: The W'_a do not satisfy the one-nonzero-weight-per-row constraint.

Because the set of matrices Q that meet our requirements is infinite, so is the number of $\{W'_a\}$ for each $\{W_a\}$. A network being trained without weight constraints is free to

discover virtually *any* of these $\{W'_a\}$ (the only restriction being that the mapping from
x' to **x** is the identity for the output units, because error is computed from **x'** directly,
not **x**). Thus, each solution $\{W_a\}$ that meets the formal definition of an update graph is
only a special case—with **Q** being the identity matrix—of the more general $\{W'_a\}$. Requiring SLUG to discover this particular solution can complicate learning, as we observed when
training SLUG with weight constraints.

If the connectivity restrictions on **W** that define an update graph could be generalized
to **W'**, these generalized restrictions could be applied to discover a large set of solutions
that nonetheless correspond to an update graph. Unfortunately, it does not appear that the
restrictions can be mapped in any straightforward way.

An alternative approach we have considered is to train SLUG to discover a solution
$\{W'_a\}$ which can then be decomposed into an update graph $\{W_a\}$ and a transformation
matrix **Q**. The decomposition could be attempted post hoc, but our experiments thus far
have consisted of explicitly defining **W'** in terms of **Q** and **W**. That is, an error gradient
is computed with respect to **W'**, which is then translated to gradients with respect to **Q**
and **W**.[7] In addition, the previously-described constraints on **W** are applied. Although SLUG
must still discover the update graph $\{W_a\}$, we hoped that introducing **Q** would allow alternative paths to the solution. This method did help somewhat, but unfortunately, performance did not reach the same levels as training with no weight constraints whatsoever. Figure 8 shows one solution obtained by SLUG under this training regimen.

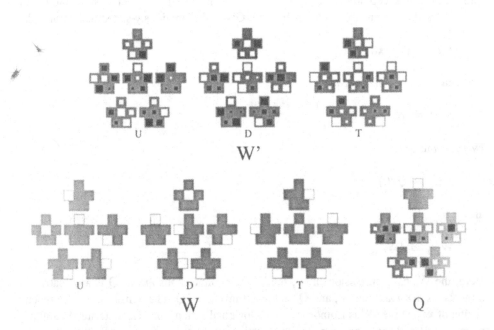

Figure 8. The three-room world. The $\{W'_a\}$ are weight matrices learned by SLUG with six units (along with
an activity vector, **x'**). Although the weights do not appear to correspond to an update graph, they in fact can
be decomposed into matrices **Q** and $\{W_a\}$ according to $W'_a = QW_aQ^*$.

7. Conclusion

The connectionist approach to the problem of inferring the structure of a finite-state environment has two fundamental problems that must be overcome if it is to be considered seriously as an alternative to the symbolic approach. First, using a random exploration strategy, SLUG has no hope of scaling to complex environments. An intelligent exploration strategy could potentially be incorporated to force the robot into states where SLUG's predictive abilities are poor. (For a related approach, see Cohn et al., 1990). Second, our greatest successes have occurred when we allowed SLUG to discover solutions that are not necessarily isomorphic to an update graph. One virtue of the update graph formalism is that it is relatively easy to interpret; the same cannot generally be said of the continuous-valued weight matrices discovered by SLUG. However, as we discussed, there is promise of developing methods for transforming the large class of formally equivalent solutions available to SLUG into the more localist update graph formalism to facilitate interpretation.

On a positive note, the connectionist approach has shown several benefits.

- Connectionist learning algorithms provide the potential of parallelism. After the outcome of a single action is observed, nearly all weights in the network are adjusted simultaneously. In contrast, the RS algorithm performs actions in order to test one or a small number of hypotheses, say, whether two particular nodes should be connected. A further example of SLUG's parallelism is that it learns the update graph structure at the same time as the appropriate unit activations, whereas the RS algorithm approaches the two tasks sequentially.

- Performance of the learning algorithm appears insensitive to prior knowledge of the number of nodes in the update graph being learned. As long as SLUG is given at least as many units as required, the presence of additional units does not impede learning. SLUG either disconnects the unnecessary units from the graph or uses multiple units to encode information redundantly. In constast, the RS algorithm requires an upper bound on the update graph complexity, and performance degrades significantly if the upper bound isn't tight.

- During learning, SLUG continually makes predictions about what sensations will result from a particular action. These predictions gradually improve with experience, and even before learning is complete, the predictions can be substantially correct. The RS algorithm cannot make predictions based on its partially constructed update graph. Although the algorithm could perhaps be modified to do so, there would be an associated cost.

- Connectionist learning algorithms are able to accommodate environments in which the sensations are somewhat unreliable. The original RS algorithm was designed for deterministic environments.

- Treating the update graph as matrices of connection strengths has suggested generalizations of the update graph formalism that don't arise from a more traditional analysis. We presented two generalizations. First, there is the fairly direct extension of allowing complementation links. Second, because SLUG is a linear system, any rank-preserving linear transform of the weight matrices will produce an equivalent system, but one that does not have the local connectivity of the update graph. Thus, one can view the Rivest and Schapire update graph formalism as one example of a much larger class of equivalent solutions that can be embodied in a connectionist network. While many of these solutions do not obey constraints imposed by a symbolic description (e.g., all-or-none links between nodes), they do yield equivalent behavior. By relaxing the symbolic constraints, the connectionist representation allows for greater flexibility in expressing potential solutions.

We emphatically do not claim that the connectionist approach supercedes the impressive work of Rivest and Schapire. However, it offers complementary strengths and an alternative conceptualization of the learning problem.

Appendix

In explaining the update graph, we informally characterized the nodes of the graph as representing sensations in the environment relative to the observer's current position. However, the nodes have a formal semantics in Rivest and Schapire's derivation of the update graph, which we present in this Appendix after first introducing a bit more terminology.

A *test* on the environment can be defined as a sequence of zero or more actions followed by a sensation. A test is performed by executing the sequence of actions from the current environmental state and then detecting the resulting sensation. For example, in the n-room world, some tests include: UU? (move up twice and sense the state of the light), UTD? (move up, toggle the light, move down, and then sense the state of the light), ? (sense the state of the light in the current room). The value of a test is the value of the resulting sensation.[8]

Certain tests are equivalent. For example, in the three-room world UU? and D? will always yield the same outcome, independent of the current environmental state. This is because moving up twice will land the robot in the same room as moving down once, so the resulting sensation will be the same. The tests UUUUU?, TUU?, DUTDDDD? are also equivalent to UU? and D?. The set of equivalent tests defines an *equivalence class*. Rivest and Schapire call the number of equivalence classes the *diversity* of the environment. The n-room world has a diversity of $2n$, arising from there being n lights that can be sensed, either in their current state or toggled. Each node in the update graph corresponds to one test equivalence class.

The directed, labeled links of the update graph arise from relations among equivalence classes: There is a link from node α to node β labeled with *action* if the test represented by α is equivalent to (i.e., will always yield the same result as) executing *action* followed by test β. For instance, there is a link from the node representing the class containing T? (which we have indicated as $\overline{\text{CUR}}$ in Figure 3e) to the node representing the class containing UT? (which we have indicated as $\overline{\text{UP}}$ in Figure 3e), and this link is labeled D because executing the action D followed by the test UT? is equivalent to executing the test T? (the U and D actions cancel).

The boolean variable associated with each node represents the truth value of the corresponding test given the current global state of the environment. If the value of each node is known prior to performing an action, the value of each node following the action is easily determined by propagating values along the links. Consider again the two nodes α and β with a directed link from α to β labeled by *action*. Because test α is equivalent to *action* followed by test β, the value of β after performing *action* is simply the value of α prior to the action.

Based on the semantics of nodes and links, it is clear why each node has exactly one incoming link for each action. If a node β received projections from two nodes, α_1 and α_2 for some *action*, this would imply that the equivalence classes α_1 and α_2 both contained the test consisting of *action* followed by test β. If the two equivalence classes contain the same test, they must represent the same equivalence class, and hence will be collapsed together.

Acknowledgments

Our thanks to Liz Jessup, Clayton Lewis, Rob Schapire, Paul Smolensky, Rich Sutton, and Dave Touretzky for helpful discussions and comments. This work was supported by NSF Presidential Young Investigator award IRI-9058450, grant 90-21 from the James S. McDonnell Foundation, and DEC external research grant 1250 to the first author; grant 87-2-36 from the Sloan Foundation to Geoffrey Hinton; and the Air Force Office of Scientific Research, Bolling AFB, under Grant AFOSR-89-0526, and the National Science Foundation under Grant ECS-8912623 to Andrew Barto.

Notes

1. The disadvantage of the update graph is that in degenerate, completely unstructured environments, the size of the update graph can be exponentially larger than the size of the FSA.
2. A consequence of this substitution is that error should not be back propagated from time t to output units at times $t - 1$, $t - 2$, etc. It is not sensible to adjust the response properties of output units at time, say, $t - 1$ to achieve the correct response at time t because their appropriate activation levels have already been established by the sensations at time $t - 1$.
3. Keeping the original value of $x(t - \tau)$ is a somewhat arbitrary choice. Consistency can be achieved by propagating *any* value of $x(t - \tau)$ forward in time, and there is no strong reason for believing $x(t - \tau)$ is the appropriate value. We thus suggest two alternative schemes, but have not yet tested them. First, we might select $x(t - \tau)$ such that the new $x(t - i)$, $i = 0 \ldots \tau - 1$, are as close as possible to the old values. Second, we might select $x(t - \tau)$ such that the output units produce as close to the correct values as possible. Both these schemes require the computation-intensive operation of finding a least squares solution to a set of linear equations.
4. The definitive connectionist light bulb joke (courtesy of Thomas Mastaglio):

   Q: How many connectionist networks does it take to change a light bulb?
   A: Only one, but it needs about 6,000 trials.

5. We thank Rob Schapire for providing us with the latest results from his work.
6. Just as connectionist simulations require a bit of voodoo in setting learning rates, the RS algorithm has its own set of adjustable parameters that influence performance. One of us (JB) experimented with the RS algorithm, and without expertise in parameter tweaking, was unable to obtain performance in the same range as the measures reported in Table 2.
7. This is not a simple matter due to the fact that $\mathbf{W}'$ is composed of $\mathbf{Q}^*$ as well as $\mathbf{Q}$, and the $\mathbf{Q}^*$ gradient must be transformed into a $\mathbf{Q}$ gradient. Consequently, we constrained $\mathbf{Q}$ to be an orthogonal matrix. For orthogonal matrices, $\mathbf{Q}^{-1} = \mathbf{Q}^T$, which trivializes the mapping from $\mathbf{Q}^*$ gradients to $\mathbf{Q}$ gradients.
8. In the n-room world, there is only one sensation—the state of the light; thus, each test ends by evaluating this sensation. In environments having multiple sensations, tests can end with different sensations.

References

Cohn, D., Atlas, L., Ladner, R., Marks II, R., El-sharkawi, M., Aggoune, M., & Park, D. (1990). Training connectionist networks with queries and selective sampling. In D.S. Touretzky (Ed.), *Advances in neural information processing systems 2* (pp. 566-573). San Mateo, CA: Morgan Kaufmann.

Elman, J.L. (1988). *Finding structure in time* (CRL Technical Report 8801). La Jolla: University of California, San Diego, Center for Research in Language.

Mozer, M.C. (1989). A focused back-propagation algorithm for temporal pattern recognition. *Complex Systems, 3,* 349-381.

Rivest, R.L., & Schapire, R.E. (1987). Diversity-based inference of finite automata. In *Proceedings of the Twenty-Eighth Annual Symposium on Foundations of Computer Science* (pp. 78-87).

Rivest, R.L., & Schapire, R.E. (1987). A new approach to unsupervised learning in deterministic environments. In P. Langley (Ed.), *Proceedings of the Fourth International Workshop on Machine Learning* (pp. 364-375).

Rumelhart, D.E., Hinton, G.E., & Williams, R.J. (1986). Learning internal representations by error propagation. In D.E. Rumelhart & J.L. McClelland (Eds.), *Parallel distributed processing: Explorations in the microstructure of cognition. Volume I: Foundations* (pp. 318-362). Cambridge, MA: MIT Press/Bradford Books.

Schapire, R.E. (1988). *Diversity-based inference of finite automata.* Unpublished master's thesis, Massachusetts Institute of Technology, Cambridge, MA.

Servan-Schreiber, D., Cleeremans, A., & McClelland, J.L. (1988). *Encoding sequential structure in simple recurrent networks* (Technical Report CMU-CS-88-183). Pittsburgh, PA: Carnegie-Mellon University, Department of Computer Science.

Machine Learning, 7, 161–193 (1991)

Graded State Machines: The Representation of Temporal Contingencies in Simple Recurrent Networks

DAVID SERVAN-SCHREIBER, AXEL CLEEREMANS AND JAMES L. MCCLELLAND
School of Computer Science and Department of Psychology, Carnegie Mellon University

Abstract. We explore a network architecture introduced by Elman (1990) for predicting successive elements of a sequence. The network uses the pattern of activation over a set of hidden units from time-step t-1, together with element t, to predict element t + 1. When the network is trained with strings from a particular finite-state grammar, it can learn to be a perfect finite-state recognizer for the grammar. When the net has a minimal number of hidden units, patterns on the hidden units come to correspond to the nodes of the grammar; however, this correspondence is not necessary for the network to act as a perfect finite-state recognizer. Next, we provide a detailed analysis of how the network acquires its internal representations. We show that the network progressively encodes more and more temporal context by means of a probability analysis. Finally, we explore the conditions under which the network can carry information about distant sequential contingencies across intervening elements to distant elements. Such information is maintained with relative ease if it is relevant at each intermediate step; it tends to be lost when intervening elements do not depend on it. At first glance this may suggest that such networks are not relevant to natural language, in which dependencies may span indefinite distances. However, embeddings in natural language are not completely independent of earlier information. The final simulation shows that long distance sequential contingencies can be encoded by the network even if only subtle statistical properties of embedded strings depend on the early information. The network encodes long-distance dependencies by *shading* internal representations that are responsible for processing common embeddings in otherwise different sequences. This ability to represent simultaneously similarities and differences between several sequences relies on the graded nature of representations used by the network, which contrast with the finite states of traditional automata. For this reason, the network and other similar architectures may be called *Graded State Machines*.

Keywords. Graded state machines, finite state automata, recurrent networks, temporal contingencies, prediction task

1. Introduction

As language abundantly illustrates, the meaning of individual events in a stream—such as words in a sentence—is often determined by preceding events in the sequence, which provide a context. The word 'ball' is interpreted differently in "The countess threw the ball" and in "The pitcher threw the ball." Similarly, goal-directed behavior and planning are characterized by coordination of behaviors over long sequences of input-output pairings, again implying that goals and plans act as a context for the interpretation and generation of individual events.

The similarity-based style of processing in connectionist models provides natural primitives to implement the role of context in the selection of meaning and actions. However, most connectionist models of sequence processing present all cues of a sequence in parallel and often assume a fixed length for the sequence (e.g., Cottrell, 1985; Fanty, 1985; Selman, 1985; Sejnowski & Rosenberg, 1987; Hanson and Kegl, 1987). Typically, these models use a pool of input units for the event present at time **t**, another pool for event **t + 1**, and so on, in what is often called a 'moving window' paradigm. As Elman (1990) points out, such implementations are not psychologically satisfying, and they are also computationally wasteful since some unused pools of units must be kept available for the rare occasions when the longest sequences are presented.

Some connectionist architectures have specifically addressed the problem of learning and representing the information contained in sequences in more elegant ways. Jordan (1986) described a network in which the output associated to each state was fed back and blended with the input representing the next sate over a set of 'state units' (Figure 1).

After several steps of processing, the pattern present on the input units is characteristic of the particular sequence of states that the network has traversed. With sequences of increasing length, the network has more difficulty discriminating on the basis of the first cues presented, but the architecture does not rigidly constrain the length of input sequences. However, while such a network learns *how to use* the representation of successive states, it does not discover a representation for the sequence.

Elman (1990) has introduced an architecture—which we call a simple recurrent network (SRN)—that has the potential to master an infinite corpus of sequences with the limited means of a learning procedure that is *completely local in time* (Figure 2). In the SRN, the hidden unit layer is allowed to feed back on itself, so that the intermediate results of processing at time **t − 1** can influence the intermediate results of processing at time **t**. In practice, the simple recurrent network is implemented by copying the pattern of activation on the hidden units onto a set of 'context units' which feed into the hidden layer along with the input units. These context units are comparable to Jordan's state units.

In Elman's simple recurrent networks, the set of context units provides the system with memory in the form of a trace of processing at the previous time slice. As Rumelhart, Hinton and Williams (1986) have pointed out, the pattern of activation on the hidden units

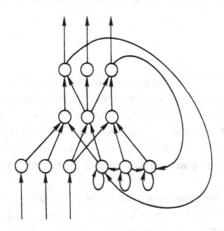

Figure 1. The Jordan (1986) Sequential Network.

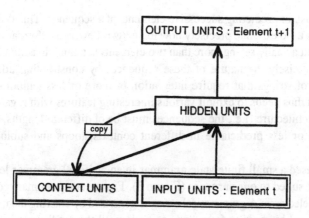

Figure 2. The Simple Recurrent Network. Each box represents a pool of units and each forward arrow represents a complete set of trainable connections from each sending unit to each receiving unit in the next pool. The backward arrow from the hidden layer to the context layer denotes a copy operation.

corresponds to an 'encoding' or 'internal representation' of the input pattern. By the nature of back-propagation, such representations correspond to the input pattern partially processed into features relevant to the task (e.g., Hinton, McClelland & Rumelhart, 1986). In the recurrent networks, internal representations encode not only the prior event but also relevant aspects of the representation that was constructed in predicting the prior event from its predecessor. When fed back as input, these representations could provide information that allows the network to maintain prediction-relevant features of an entire sequence.

In this study, we show that the SRN can learn to mimic closely a finite state automaton (FSA), both in its behavior and in its state representations. In particular, we show that it can learn to process an *infinite* corpus of strings based on experience with a *finite* set of training exemplars. We then explore the capacity of this architecture to recognize and use non-local contingencies between elements of a sequence that cannot be represented conveniently in a traditional finite state automaton. We show that the SRN encodes long-distance dependencies by *shading* internal representations that are responsible for processing common embeddings in otherwise different sequences. This ability to represent simultaneously similarities and differences between sequences in the same state of activation relies on the graded nature of representations used by the network, which contrast with the finite states of traditional automata. For this reason, we suggest that the SRN and other similar architectures may be exemplars of a new class of automata, one that we may call *Graded State Machines*.

2. Learning a finite state grammar

2.1. Material and task

In our first experiment, we asked whether the network cold learn the contingencies implied by a small finite state grammar. As in all of the following explorations, the network is assigned the task of predicting successive elements of a sequence. This task is interesting because it allows us to examine precisely how the network extracts information about whole sequences without actually seeing more than two elements at a time. In addition, it is possible to manipulate precisely the nature of these sequences by constructing different training and testing sets of strings that require integration of more or less temporal information. The stimulus set thus needs to exhibit various interesting features with regard to the potentialities of the architecture (i.e., the sequences must be of different lengths, their elements should be more or less predictable in different contexts, loops and subloops should be allowed, etc.).

Reber (1976) used a small finite-state grammar in an artificial grammar learning experiment that is well suited to our purposes (Figure 3). Finite-state grammars consist of nodes connected by labeled arcs. A grammatical string is generated by entering the network through the 'start' node and by moving from node to node until the 'end' node is reached. Each transition from one node to another produces the letter corresponding to the label of the arc linking these two nodes. Examples of strings that can be generated by the above grammar are: 'TXS', 'PTVV', 'TSXXTVPS'.

The difficulty in mastering the prediction task when letters of a string are presented individually is that two instances of the same letter may lead to different nodes and therefore different predictions about its successors. In order to perform the task adequately, it is thus necessary for the network to encode more than just the identity of the current letter.

2.2. Network architecture

As illustrated in Figure 4, the network has a three-layer architecture. The input layer consists of two pools of units. The first pool is called the context pool, and its units are used to represent the temporal context by holding a copy of the hidden units' activation level at the previous time slice (note that this is strictly equivalent to a fully connected feedback loop on the hidden layer). The second pool of input units represents the current element

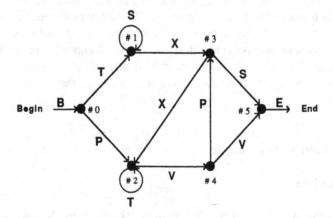

Figure 3. The finite-state grammar used by Reber (1976).

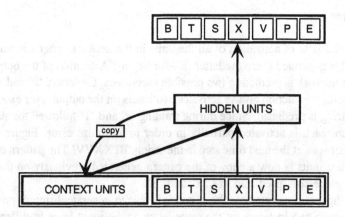

Figure 4. General architecture of the network.

of the string. On each trial, the network is presented with an element of the string, and is supposed to produce the next element on the output layer. In both the input and the output layers, letters are represented by the activation of a single unit. Five units therefore code for the five different possible letters in each of these two layers. In addition, two units code for *begin* and *end* bits. These two bits are needed so that the network can be trained to predict the first element and the end of a string (although only one *transition* bit is strictly necessary). In this first experiment, the number of hidden units was set to 3. Other values will be reported as appropriate.

2.3. Coding of the strings

A string of **n** letters is coded as a series of **n + 1** training patterns. Each pattern consists of two input vectors and one target vector. The target vector is a seven-bit vector representing element **t + 1** of the string. The two input vectors are:

- A three-bit vector representing the activation of the hidden units at time **t − 1**, and
- A seven-bit vector representing element **t** of the string.

2.4. Training

On each of 60,000 training trials, a string was generated from the grammar, starting with the 'B.' Successive arcs were then selected randomly from the two possible continuations, with a probability of 0.5. Each letter was then presented sequentially to the network. The activations of the context units were reset to 0 at the beginning of each string. After each letter, the error between the network's prediction and the *actual successor* specified by the string was computed and back-propagated. The 60,000 randomly generated strings ranged from 3 to 30 letters (mean: 7, sd: 3.3)[1].

2.5. Performance

Figure 5 shows the state of activation of all the units in the network, after training, when the start symbol is presented (here the letter 'B'—for begin). Activation of the output units indicate that the network is predicting two possible successors, the letters 'P' and 'T.' Note that the best possible prediction always activates two letters on the output layer except when the end of the string is predicted. Since during training 'P' and 'T' followed the start symbol equally often, each is activated partially in order to minimize error. Figure 6 shows the state of the network at the next time step in the string 'BTXXVV.' The pattern of activation on the context units is now a copy of the pattern generated previously on the hidden layer. The two successors predicted are 'X' and 'S.'

The next two figures illustrate how the network is able to generate different predictions when presented with two instances of the same letter on the input layer in different contexts. In Figure 7a, when the letter 'X' immediately follows 'T,' the network predicts again 'S' and 'X' appropriately. However, as Figure 7b shows, when a second 'X' follows, the prediction changes radically as the network now expects 'T' or 'V.' Note that if the network were not provided with a copy of the previous pattern of activation on the hidden layer, it would activate the four possible successors of the letter 'X' in both cases.

```
String     B t x x v v
           B   T   S   P   X   V   E
Output     00 53  00  38  01  02  00
Hidden             01  00  10
Context            00  00  00
           B   T   S   P   X   V   E
Input      100 00  00  00  00  00  00
```

Figure 5. State of the network after presentation of the 'Begin' symbol (following training). Activation values are internally in the range 0 to 1.0 and are displayed on a scale from 0 to 100. The capitalized bold letter indicates which letter is currently being presented on the input layer.

```
String     b T x x v v
           B   T   S   P   X   V   E
Output     00 01  39  00  56  00  00
Hidden             84  00  28
Context            01  00  10
           B   T   S   P   X   V   E
Input      00 100  00  00  00  00  00
```

Figure 6. State of the network after presentation of an initial 'T.' Note that the activation pattern on the context layer is identical to the activation pattern on the hidden layer at the previous time step.

```
String      b t X x v v
            B  T  S  P  X  V  E
Output      00 04 44 00 37 07 00
Hidden         74 00 93
Context        84 00 28
            B  T  S  P  X  V  E
Input       00 00 00 00 100 00 00
```

a

```
String      b t x X v v
            B  T  S  P  X  V  E
Output      00 50 01 01 00 55 00
Hidden         06 09 99
Context        74 00 93
            B  T  S  P  X  V  E
Input       00 00 00 00 100 00 00
```

b

Figure 7. a) State of the network after presentation of the first 'X.' b) State of the network after presentation of the second 'X.'

In order to test whether the network would generate similarly good predictions after every letter of any grammatical string, we tested its behavior on 20,000 strings derived randomly from the grammar. A prediction was considered accurate if, for every letter in a given string, activation of its successor was above 0.3. If this criterion was not met, presentation of the string was stopped and the string was considered 'rejected.' With this criterion, the network correctly 'accepted' all of the 20,000 strings presented.

We also verified that the network did not accept ungrammatical strings. We presented the network with 130,000 strings generated from the same pool of letters but in a random manner—i.e., mostly 'non-grammatical.' During this test, the network is first presented with the 'B' and one of the five letters or 'E' is then selected at random as a successor. If that letter is predicted by the network as a legal successor (i.e., activation is above 0.3 for the corresponding unit), it is then presented to the input layer on the next time step, and another letter is drawn at random as its successor. This procedure is repeated as long as each letter is predicted as a legal successor until 'E' is selected as the next letter. The procedure is interrupted as soon as the actual successor generated by the random procedure is *not* predicted by the network, and the string of letters is then considered 'rejected.' As in the previous test, the string is considered 'accepted' if all its letters have been predicted as possible continuations up to 'E.' Of the 130,000 strings, 0.2% (260) happened to be grammatical, and 99.7% were non-grammatical. The network performed flawlessly, accepting all the grammatical strings and rejecting all the others. In other words, for all non-grammatical strings, when the first non-grammatical letter was presented to the network its activation on the output layer at the previous step was less than 0.3 (i.e., it was *not* predicted as a successor of the previous—grammatically acceptable—letter).

Finally, we presented the network with several extremely long strings such as:

'BTSSSSSSSSSSSSSSSSSSSSSSSSXXVPXVPXVPXVPXVPXVPXVPXVPXVPXVPXTT
TTTTTTTTTTTTTTTTTTTTTTTTTTTTTTVPXVPXVPXVPXVPXVPXVPS'

and observed that, at every step, the network correctly predicted both legal successors and no others.

Note that it is possible for a network with more hidden units to reach this performance criterion with much less training. For example, a network with 15 hidden units reached criterion after 20,000 strings were presented. However, activation values on the output layer are not as clearly contrasted when training is less extensive. Also, the selection of a threshold of 0.3 is not completely arbitrary. The activation of output units is related to the frequency with which a particular letter appears as the successor of a given sequence. In the training set used here, this probability is 0.5. The activation of a legal successor would then be expected to be 0.5.[2] However, because of the use of a momentum term in the back propagation learning procedure, the activation of correct output units following training was occasionally below 0.5—sometimes as low as 0.3.

2.6. Analysis of internal representations

Obviously, in order to perform accurately, the network takes advantage of the representations that have developed on the hidden units which are copied back onto the context layer. At any point in the sequence, these patterns must somehow encode the position of the current input in the grammar on which the network was trained. One approach to understanding how the network uses these patterns of activation is to perform a cluster analysis. We recorded the patterns of activation on the hidden units following the presentation of each letter in a small random set of grammatical strings. The matrix of Euclidean distances between each pair of vectors of activation served as input to a cluster analysis program.[3] The graphical result of this analysis is presented in Figure 8A. Each leaf in the tree corresponds to a particular string, and the capitalized letter in that string indicates which letter has just been presented. For example, if the leaf is identified as 'pvPs,' 'P' is the current letter and its predecessors were 'P' and 'V' (the correct prediction would thus be 'X' or 'S').

From the figure, it is clear that activation patterns are grouped according to the different nodes in the finite state grammar; all the patterns that produce a similar prediction are grouped together, independently of the current letter. This grouping by similar predictions is apparent in Figure 8b, which represents an enlargement of the bottom cluster (cluster 5) of Figure 8a. One can see that this cluster groups patterns that result in the activation of the "End" unit: All the strings corresponding to these patterns end in 'V' or 'S' and lead to node #5 of the grammar, out of which "End" is the only possible successor. Therefore, when one of the hidden layer patterns is copied back onto the context layer, the network is provided with information about the *current node*. That information is combined with input representing the *current letter* to produce a pattern on the hidden layer that is a representation of the *next node*. To a degree of approximation, the recurrent network behaves exactly like the finite state automaton defined by the grammar. It does not use a stack or registers to provide contextual information but relies instead on simple state transitions, just like a finite state machine. Indeed, the network's perfect performance on randomly generated grammatical and non-grammatical strings shows that it can be used as a finite state recognizer.

However, a closer look at the cluster analysis reveals that within a cluster corresponding to a particular node, patterns are further divided according to the path traversed before that node. For example, an examination of Figure 8b reveals that patterns ending by 'VV,' 'PS' and 'SXS' endings have been grouped separately by the analysis: they are more similar to each other than to the abstract prototypical pattern that would characterize the corresponding "node."[4] We can illustrate the behavior of the network with a specific example. When the first letter of the string 'BTX' is presented, the initial pattern on context units corresponds to node 0. This pattern together with the letter 'T' generates a hidden layer pattern corresponding to node 1. When that pattern is copied onto the context layer and the letter 'X' is presented, a new pattern corresponding to node 3 is produced on the hidden layer, and this pattern is in turn copied on the context units. If the network behaved *exactly* like a finite state automaton, the exact same patterns would be used during processing of the other strings 'BTSX' and 'BTSSX.' That behavior would be adequately captured by the transition network shown in Figure 9. However, since the cluster analysis shows that slightly different patterns are produced by the substrings 'BT,' 'BTS' and 'BTSS,' Figure 10 is a more accurate description of the network's state transitions. As states 1, 1' and 1" on the one hand and 3, 3' and 3" on the other are nevertheless very similar to each other, the finite state machine that the network implements can be said to approximate the idealization of a finite state automaton corresponding exactly to the grammar underlying the exemplars on which it has been trained.

However, we should point out that the close correspondence between representations and function obtained for the recurrent network with three hidden units is rather the exception than the rule. With only three hidden units, representational resources are so scarce that backpropagation forces the network to develop representations that yield a prediction on the basis of the current node alone, ignoring contributions from the path. This situation precludes the development of different—redundant—representations for a particular node that typically occurs with larger numbers of hidden units. When redundant representations do develop, the network's behavior still converges to the theoretical finite state automaton—in the sense that it can still be used as a perfect finite state recognizer for strings generated from the corresponding grammar—but internal representations do not correspond to that idealization. Figure 11 shows the cluster analysis obtained from a network with 15 hidden units after training on the same task. Only nodes 4 and 5 of the grammar seem to be

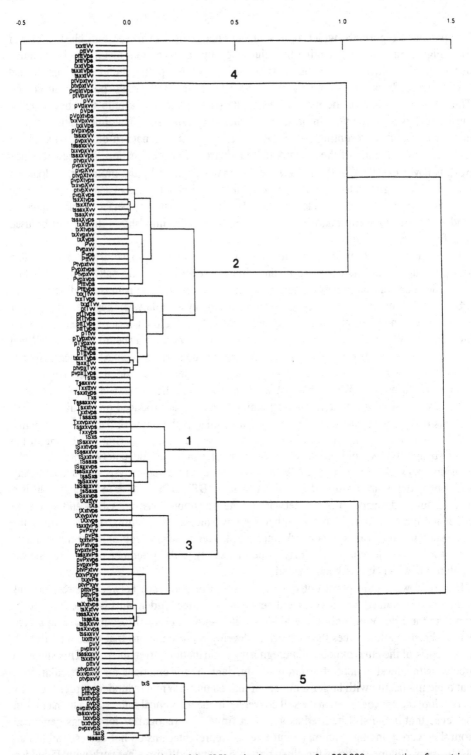

Figure 8A. Hierarchical Cluster Analysis of the H.U. activation patterns after 200,000 presentations from strings generated at random according to the Reber grammar (Three hidden units).

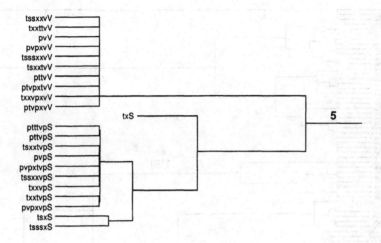

Figure 8B. An enlarged portion of Figure 8A, representing the bottom cluster (cluster 5). The proportions of the original figure have not necessarily been respected in this enlargement.

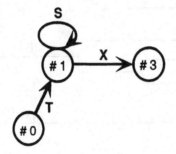

Figure 9. A transition network corresponding to the upper-left part of Reber's finite-state grammar.

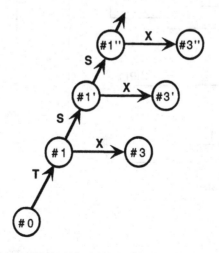

Figure 10. A transition network illustrating the network's true behavior.

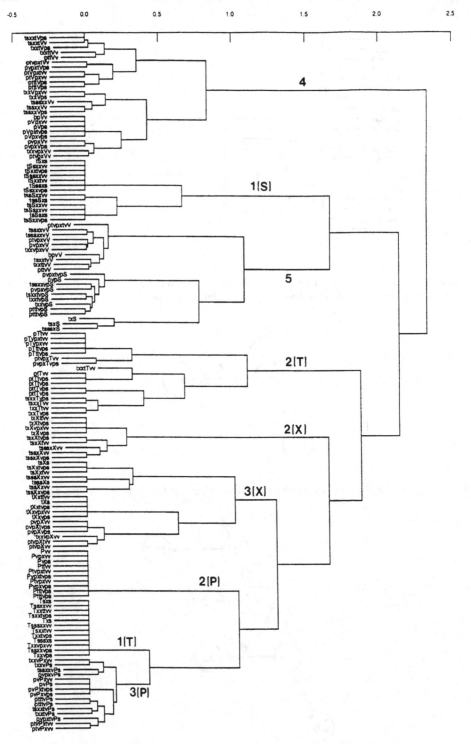

Figure 11. Hierarchical Cluster Analysis of the H.U. activation patterns after 200,000 presentations from strings generated at random according to the Reber grammar (Fifteen hidden units).

represented by a unique 'prototype' on the hidden layer. Clusters corresponding to nodes 1, 2 and 3 are divided according to the preceding arc. Information about arcs is not relevant to the prediction task and the different clusters corresponding to the a single node play a redundant role.

Finally, preventing the development of redundant representations may also produce adverse effects. For example, in the Reber grammar, predictions following nodes 1 and 3 are identical ('X or S'). With some random sets of weights and training sequences, networks with only three hidden units occasionally develop almost identical representations for nodes 1 and 3, and are therefore unable to differentiate the first from the second 'X' in a string.

In the next section we examine a different type of training environment, one in which information about the path traversed becomes relevant to the prediction task.

3. Discovering and using path information

The previous section has shown that simple recurrent networks can learn to encode the nodes of the grammar used to generate strings in the training set. However, this training material does not require information about arcs or sequences of arcs—the 'path'—to be maintained. How does the network's performance adjust as the training material involves more complex and subtle temporal contingencies? We examine this question in the following section, using a training set that places many additional constraints on the prediction task.

3.1. Material

The set of strings that can be generated from the grammar is finite for a given length. For lengths 3 to 8, this amounts to 43 grammatical strings. The 21 strings shown in Figure 12 were selected and served as training set. The remaining 22 strings can be used to test generalization.

The selected set of strings has a number of interesting properties with regard to exploring the network's performance on subtle temporal contingencies:

- As in the previous task, identical letters occur at different points in each string, and lead to different predictions about the identity of the successor. No stable prediction is therefore associated with any particular letter, and it is thus necessary to encode the position, or the node of the grammar.

TSXS	TXS	TSSSXS
TSSXXVV	TSSSXXVV	TXXVPXVV
TXXTTVV	TSXXTVV	TSSXXVPS
TSXXTVPS	TXXTVPS	TXXVPS
PVV	PTTVV	PTVPXVV
PVPXVV	PTVPXTVV	PVPXVPS
PTVPS	PVPXTVPS	PTTTVPS

Figure 12. The 21 grammatical strings of length 3 to 8.

- In this limited training set, length places additional constraints on the encoding because the possible predictions associated with a particular node in the grammar are dependent on the length of the sequence. The set of possible letters that follow a particular node depends on how many letters have already been presented. For example, following the sequence 'TXX' both 'T' and 'V' are legal successors. However, following the sequence 'TXXVPX,' 'X' is the sixth letter and only 'V' would be a legal successor. This information must therefore also be somehow represented during processing.
- Subpatterns occurring in the strings are not all associated with their possible successors equally often. Accurate predictions therefore require that information about the identity of the letters that have already been presented be maintained in the system, i.e., the system must be sensitive to the frequency distribution of subpatterns in the training set. This amounts to encoding the full path that has been traversed in the grammar.

These features of the limited training set obviously make the prediction task much more complex than in the previous simulation.

3.2. Network architecture

The same general network architecture was used for this set of simulations. The number of hidden units was arbitrarily set to 15.

3.3. Performance

The network was trained on the 21 different sequences (a total of 130 patterns) until the total sum squared error (*tss*) reached a plateau with no further improvements. This point was reached after 2000 epochs and *tss* was 50. Note that *tss* cannot be driven much below this value, since most partial sequences of letters are compatible with 2 different successors. At this point, the network correctly predicts the possible successors of each letter, and distinguishes between different occurrences of the same letter—like it did in the simulation described previously. However, the network's performance makes it obvious that many additional constraints specific to the limited training set have been encoded. Figure 13a shows that the network expects a 'T' or a 'V' after a first presentation of the second 'X' in the grammar.

Contrast these predictions with those illustrated in Figure 13b, which shows the state of the network after a *second* presentation of the second 'X': Although the same node in the grammar has been reached, and 'T' and 'V' are again possible alternatives, the network now predicts only 'V.'

Thus, the network has successfully learned that an 'X' occurring late in the sequence is never followed by a 'T'—a fact which derives directly from the maximum length constraint of 8 letters.

It could be argued that that network simply learned that when 'X' is preceded by 'P' it cannot be followed by 'T', and thus relies only on the preceding letter to make that distinction. However, the story is more complicated than this.

```
String     b t x X v p x v v
           B   T   S   P   X   V   E
Output     00  49  00  00  00  50  00
Hidden     27 89 02 16 99 43 01 06 04 18 99 81 95 18 01
Context    01 18 00 41 95 01 60 59 05 06 84 99 19 05 00
           B   T   S   P   X   V   E
Input      00  00  00  00 100 00  00
```

a

```
String     b t x x v p X v v
           B   T   S   P   X   V   E
Output     00  03  00  00  00  95  00
Hidden     00 85 03 85 31 00 72 19 31 03 93 99 61 05 00
Context    01 07 05 90 93 04 00 10 71 40 99 16 90 05 82
           B   T   S   P   X   V   E
Input      00  00  00  00 100 00  00
```

b

Figure 13. a) State of the network after presentation of the second 'X.' b) State of the network after a second presentation of the second 'X.'

In the following two cases, the network is presented with the first occurrence of the letter 'V.' In the first case, 'V' is preceded by the sequence 'tssxx,' while in the second case, it is preceded by 'tsssxx.' The difference of a single 'S' in the sequence—which occurred 5 presentations before—results in markedly different predictions when 'V' is presented (Figures 14a and 14b).

The difference in predictions can be traced again to the length constraint imposed on the strings in the limited training set. In the second case, the string spans a total of 7 letters when 'V' is presented, and the only alternative compatible with the length constraint is a second 'V' and the end of the string. This is not true in the first case, in which both 'VV' and 'VPS' are possible endings.

Thus, it seems that the representation developed on the context units encodes more than the immediate context—the pattern of activation could include a full representation of the path traversed so far. Alternatively, it could be hypothesized that the context units encode only the preceding letter and a counter of how many letters have been presented.

String	b t s s x x **V** v													
	B	T	S	P	X	V	E							
Output	00	00	00	54	00	48	00							
Hidden	44	98	30	84	99	82	00	47	00	09	41	98	13	02	00
Context	89	90	01	01	99	70	01	03	02	10	99	95	85	21	00
	B	T	S	P	X	V	E							
Input	00	00	00	00	00	100	00							

a

String	b t s s s x x **V** v													
	B	T	S	P	X	V	E							
Output	00	00	00	02	00	97	00							
Hidden	56	99	48	93	99	85	00	22	00	10	77	97	30	03	00
Context	54	67	01	04	99	59	07	09	01	06	98	97	72	16	00
	B	T	S	P	X	V	E							
Input	00	00	00	00	00	100	00							

b

Figure 14. Two presentations of the first 'V,' with slightly different paths.

In order to understand better the kind of representations that encode sequential context we performed a cluster analysis on all the hidden unit patterns evoked by each sequence. Each letter of each sequence was presented to the network and the corresponding pattern of activation on the hidden layer was recorded. The Euclidean distance between each pair of patterns was computed and the matrix of all distances was provided as input to a cluster analysis program.

The resulting analysis is shown in Figure 15. We labeled the arcs according to the letter being presented (the 'current letter') and its position in the Reber grammar. Thus 'V_1' refers to the first 'V' in the grammar and 'V_2' to the second 'V' which immediately precedes the end of the string. 'Early' and 'Late' refer to whether the letter occurred early or late in the sequence (for example in 'PT. .' 'T_2' occurs early; in 'PVPXT. .' it occurs late). Finally, in the left margin we indicated what predictions the corresponding patterns yield on the output layer (e.g., the hidden unit pattern generated by 'B' predicts 'T' or 'P').

From the figure, it can be seen that the patterns are grouped according to three distinct principles: (1) according to similar predictions, (2) according to similar letters presented

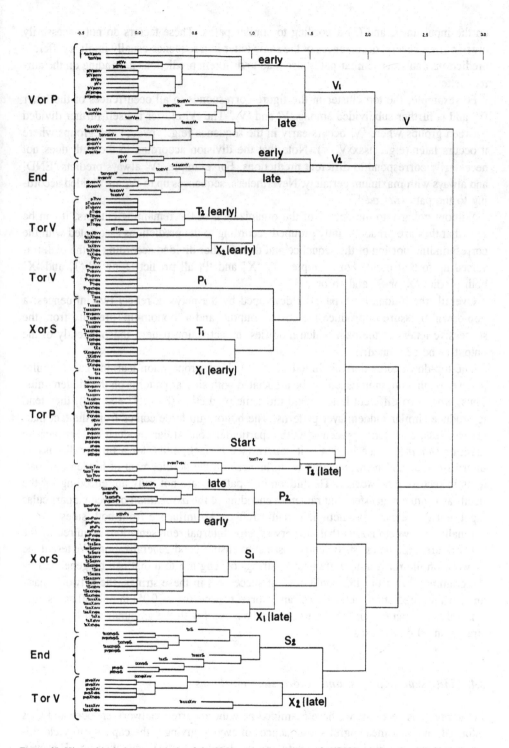

Figure 15. Hierarchical cluster analysis of the H.U. activation patters after 2000 epochs of training on the set of 21 strings.

on the input units, and (3) according to similar paths. These factors do not necessarily overlap since several occurrences of the same letter in a sequence usually implies different predictions and since similar paths also lead to different predictions depending on the current letter.

For example, the top cluster in the figure corresponds to all occurrences of the letter 'V' and is further subdivided among 'V_1^i' and 'V_2.' The 'V_1^i' cluster is itself further divided between groups where 'V_1^i' occurs early in the sequence (e.g., 'pV...') and groups where it occurs later (e.g., 'tssxxV...'). Note that the division according to the path does not necessarily correspond to different predictions. For example, 'V_2^i' always predicts 'END' and always with maximum certainty. Nevertheless, sequences up to 'V_2^i' are divided according to the path traversed.

Without going into the details of the organization of the remaining clusters, it can be seen that they are predominantly grouped according to the predictions associated with the corresponding portion of the sequence and then further divided according to the path traversed up to that point. For example, 'T_2^i' 'X_2^i' and 'P_1^i' all predict 'T or V,' 'T_1^i' and 'X_1^i' both predict 'X or S,' and so on.

Overall, the hidden units patterns developed by the network reflect two influences: a 'top-down' pressure to produce the correct ouptut, and a 'bottom-up' pressure from the successive letters in the path which modifies the activation pattern independently of the output to be generated.

The top-down force derives directly from the back-propagation learning rule. Similar patterns on the output units tend to be associated with similar patterns on the hidden units. Thus, when two different letters yield the same prediction (e.g., 'T_1^i' and 'X_1^i'), they tend to produce similar hidden layer patterns. The bottom-up force comes from the fact that, nevertheless, each letter presented with a particular context can produce a characteristic mark or *shading* on the hidden unit pattern (see Pollack, 1989, for a further discussion of error-driven and recurrence-driven influences on the development of hidden unit patterns in recurrent networks). The hidden unit patterns are not truly an 'encoding' of the input, as is often suggested, but rather an encoding of the *association* between a particular input and the relevant prediction. It really reflects an influence from both sides.

Finally, it is worth noting that the very specific internal representations acquired by the network are nonetheless sufficiently abstract to ensure good generalization. We tested the network on the remaining untrained 22 strings of length 3 to 8 that can be generated by the grammar. Over the 165 predictions of successors in these strings, the network made an incorrect prediction (activation of an incorrect successor > 0.05) in only 10 cases, and it failed to predict one of two continuations consistent with the grammar and length constraints in 10 other cases.

3.4. Finite state automata and graded state machines

In the previous sections, we have examined how the recurrent network encodes and uses information about meaningful subsequences of events, giving it the capacity to yield different outputs according to some specific traversed path or to the length of strings. However, the network does not use a separate and explicit representation for non-local properties of the strings such as length. It only learns to associate different predictions to a subset of states; those that are associated with a more restricted choice of successors. Again there are not stacks or registers, and each different prediction is associated to a specific state on the context units. In that sense, the recurrent network that has learned to master this task still behaves like a finite-state machine, although the training set involves non-local constraints that could only be encoded in a very cumbersome way in a finite-state *grammar*.

We usually do not think of finite state automata as capable of encoding non-local information such as length of a sequence. Yet, finite state machines have in principle the same computational power as a Turing machine with a finite tape and they can be designed to respond adequately to non-local constraints. Recursive or Augmented transition networks and other Turing-equivalent automata are preferable to finite state machines because they spare memory and are modular—and therefore easier to design and modify. However, the finite state machines that the recurrent network seems to implement have properties that set them apart from their traditional counterparts:

- For tasks with an appropriate structure, recurrent networks develop their own state transition diagram, sparing this burden to the designer.
- The large amount of memory required to develop different representations for every state needed is provided by the representational power of hidden layer patterns. For example, 15 hidden units with four possible values—e.g., 0, .25, .75, 1—can support more than one billion different patterns (4^{15} = 107,374,1824).
- The network implementation remains capable of performing similarity-based processing, making it somewhat noise-tolerant (the machine does not 'jam' if it encounters an undefined state transition and it can recover as the sequence of inputs continues), and it remains able to generalize to sequences that were not part of the training set.

Because of its inherent ability to use *graded* rather than finite states, the SRN is definitely not a finite state machine of the usual kind. As we mentioned above, we have come to consider it as an exemplar of a new class of automata that we call *Graded State Machines*.

In the next section, we examine how the SRN comes to develop appropriate internal representations of the temporal context.

4. Learning

We have seen that the SRN develops and learns to use compact and effective representations of the sequences presented. These representations are sufficient to disambiguate identical cues in the presence of context, to code for length constraints and to react appropriately to atypical cases.[5] How are these representations discovered?

As we noted earlier, in an SRN, the hidden layer is presented with information about the current letter, but also—on the context layer—with an encoding of the relevant features of the previous letter. Thus, a given hidden layer pattern can come to encode information about the relevant features of two consecutive letters. When this pattern is fed back on the context layer, the new pattern of activation over the hidden units can come to encode information about three consecutive letters, and so on. In this manner, the context layer patterns can allow the network to maintain prediction-relevant features of an entire sequence.

As discussed elsewhere in more detail (Servan-Schreiber, Cleeremans & McClelland, 1988, 1989), learning progresses through three qualitatively different phases. During a first phase, the network tends to ignore the context information. This is a direct consequence of the fact that the patterns of activation on the hidden layer—and hence the context layer—are continuously changing from one epoch to the next as the weights from the input units (the letters) to the hidden layer are modified. Consequently, adjustments made to the weights from the context layer to the hidden layer are inconsistent from epoch to epoch and cancel each other. In contrast, the network is able to pick up the stable association between each *letter* and all its possible successors. For example, after only 100 epochs of training, the response pattern generated by 'S$_1$' and the corresponding output are almost identical to the pattern generated by 'S$_2$', as Figures 16a and 16b demonstrate. At the end of this phase, the network thus predicts all the successors of each letter in the grammar, independently of the *arc* to which each letter corresponds.

```
Epoch      100
String     b S s x x v p s
           B    T    S    P    X    V    E
Output     00   00   36   00   33   16   17
Hidden     45 24 47 26 36 23 55 22 22 26 22 23 30 30 33
Context    44 22 56 21 36 22 64 16 13 23 20 16 25 21 40
           B    T    S    P    X    V    E
Input      00   00  100 00   00   00   00
```

a

```
Epoch      100
String     b s s x x v p S
           B    T    S    P    X    V    E
Output     00   00   37   00   33   16   17
Hidden     45 24 47 25 36 23 56 22 21 25 21 22 29 30 32
Context    42 29 53 24 32 27 61 25 16 33 25 23 28 27 41
           B    T    S    P    X    V    E
Input      00   00  100 00   00   00   00
```

b

Figure 16. a) Hidden layer and output generated by the presentation of the *first* S in a sequence after 100 epochs of training. b) Hidden layer and output patterns generated by the presentation of the *second* S in a sequence after 100 epochs of training.

In a second phase, patterns copied on the context layer are now represented by a unique code designating which letter preceded the current letter, and the network can exploit this stability of the context information to start distinguishing between different occurrences of the same letter—different arcs in the grammar. Thus, to continue with the above example, the response elicited by the presentation of an 'S$_1$' would progressively become different from that elicited by an 'S$_2$.'

Finally, in a third phase, small differences in the context information that reflect the occurrence of previous elements can be used to differentiate position-dependent predictions resulting from length constraints. For example, the network learns to differentiate between 'tssxxV' which predicts either 'P' or 'V,' and 'tsssxxV' which predicts only 'V,' although both occurrences of 'V' correspond to the same arc in the grammar. In order to make this distinction, the pattern of activation on the context layer must be a representation of the entire path rather than simply an encoding of the previous letter.

Naturally, these three phases do not reflect sharp changes in the network's behavior over training. Rather, they are simply particular points in what is essentially a continuous process, during which the network progressively encodes increasing amounts of temporal context information to refine its predictions. It is possible to analyze this smooth progression towards better predictions by noting that these predictions converge towards the optimal conditional probabilities of observing a particular successor to the sequence presented up to that point. Ultimately, given sufficient training, the SRN's responses *would become* these optimal conditional probabilities (that is, the minima in the error function are located at those points in weight space where the activations equal the optimal conditional probabilities). This observation gives us a tool for anlayzing how the predictions change over time. Indeed, the conditional probability of observing a particular letter at any point in a sequence of inputs varies according to the number of preceding elements that have been encoded. For instance, since all letters occur twice in the grammar, a system basing its predictions on only the current element of the sequence will predict all the successors of the current letter, independently of the arc to which that element corresponds. If two elements of the sequence are encoded, the uncertainty about the next event is much reduced, since in many cases, subsequences of two letters are unique, and thus provide an unambiguous cue to its possible successors. In some other cases, subtle dependencies such as those resulting from length constraints require as much as 6 elements of temporal context to be optimally predictable.

Thus, by generating a large number of strings that have exactly the same statistical properties as those used during training, it is possible to estimate the conditional probabilities of observing each letter as the successor to each possible path of a given length. The *average* conditional probability (ACP) of observing a particular letter at every node of the grammar, after a given amount of temporal context (i.e., over all paths of a given length) can then be obtained easily by weighting each individual term appropriately. This analysis can be conducted for paths of any length, thus yielding a set of ACPs for each statistical order considered.[6] Each set of ACPs can then be used as the predictor variable in a regression analysis against the network's responses, averaged in a similar way. We would expect the ACPs based on short paths to be better predictors of the SRN's behavior early in training, and the ACPs based on longer paths to be better predictors of the SRN's behavior late in training, thus revealing the fact that, during training, the network learns to base its predictions on increasingly larger amounts of temporal context.

An SRN with fifteen hidden units was trained on the 43 strings of length 3 to 8 from the Reber grammar, in exactly the same conditions as described earlier. The network was trained for 1000 epochs, and its performance tested once before training, and every 50 epochs thereafter, for a total of 21 tests. Each test consisted of 1) freezing the connections, 2) presenting the network with the entire set of strings (a total of 329 patterns) once, and 3) recording its response to each individual input pattern. Next, the average activation of each response unit (i.e., each letter in the grammar) given 6 elements of temporal context was computed (i.e., after all paths of length 6 that are followed by that letter).

In a separate analysis, seven sets of ACPs (from order 0 to order 6) were computed in the manner described above. Each of these seven sets of ACPs was then used as the predictor variable in a regression analysis on each set of average activations produced by the network. These data are represented in Figure 17. Each point represents the percentage of variance explained in the network's behavior on a particular test by the ACPs of a particular statistical order. Points corresponding to the same set of ACPs are linked together, for a total of 7 curves, each corresponding to the ACPs of a particular order.

What the figure reveals is that the network's responses are approximating the conditional probabilities of increasingly higher statistical orders. Thus, before training, the performance of the network is best explained by the 0th order ACPs (i.e., the frequency of each letter in the training set). This is due to the fact that before training, the activations of the response units tend to be almost uniform, as do the 0th order ACPs. In the next two tests (i.e., at epoch 50 and epoch 100), the network's performance is best explained by the first-order ACPs. In other words, the network's predictions during these two tests were essentially based on paths of length 1. This point in training corresponds to the first phase of learning identified earlier, during which the network's responses do not distinguish between different occurrences of the same letter.

Soon, however, the network's performance comes to be better explained by ACPs of higher statistical orders. One can see the curves corresponding to the ACPs of order 2 and 3 progressively take over, thus indicating that the network is essentially basing its predictions on paths of length 2, then of length 3. At this point, the network has entered the second phase of learning, during which it now distinguishes between different occurrences of the same letter. Later in training, the network's behavior can be seen to be better captured by ACPs based on even longer paths, first of length 4, and finally, of length 5. Note that the network remains at that stage for a much longer period of time than for shorter ACPs. This reflects the fact that encoding longer paths is more difficult. At this point, the network has started to become sensitive to subtler dependencies such as length constraints, which require an encoding of the full path traversed so far. Finally, the curve corresponding to the ACPs of order 6 can be seen to raise steadily towards increasingly better fits, only to be achieved considerably later in training.

It is worth noting that there is a large amount of overlap between the percentage of variance explained by the different sets of ACPs. This is not surprising, since most of the sets of ACPs are partially correlated with each other. Even so, we see the successive correspondence to longer and longer temporal contingencies with more and more training.

In all the learning problems we examined so far, contingencies between elements of the sequence were relevant at each processing step. In the next section, we propose a detailed

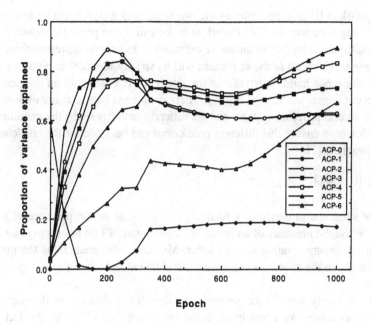

Figure 17. A graphic representation of the percentage of variance in the network's performance explained by average conditional probabilities of increasing statistical order (from 0 to 6). Each point represents the r-squared of a regression analysis using a particular set of average conditional probabilities as the predictor variable, and average activations produced by the network at a particular point in training as the dependent variable.

analysis of the constraints guiding the learning of more complex contingencies, for which information about the distant elements of the sequence to be maintained for several processing steps before they become useful.

5. Encoding non-local context

5.1. Processing loops

Consider the general problem of learning two arbitrary sequences of the same length and ending by two different letters. Under what conditions will the network be able to make a correct prediction about the nature of the last letter when presented with the penultimate letter? Obviously, the *necessary and sufficient condition* is that the internal representations associated with the penultimate letter are different (indeed, the hidden units patterns *have* to be different if different outputs are to be generated). Let us consider several different prototypical cases and verify if this condition holds:

PABC X   and   PABC V   [1]

PABC X   and   TDEF V   [2]

Clearly, problem [1] is impossible: as the two sequences are identical up to the last letter, there is simply no way for the network to make a different prediction when presented with the penultimate letter ('C' in the above example). The internal representations induced by the successive elements of the sequences will be strictly identical in both cases. Problem [2], on the other hand, is trivial, as the last letter is contingent on the penultimate letter ('X' is contingent on 'C'; 'V' on 'F'). There is no need here to maintain information available for several processing steps, and the different contexts set by the penultimate letters are sufficient to ensure that different predictions can be made for the last letter. Consider now problem [3].

PSSS P   and   TSSS T   [3]

As can be seen, the presence of a final 'T' is contingent on the presence of an initial 'T'; a final 'P' on the presence of an initial 'P.' The shared 'S's do not supply any relevant information for disambiguating the last letter. Moreover, the predictions the network is required to make in the course of processing are identical in both sequences up to the last letter.

Obviously, the only way for the network to solve this problem is to develop *different internal representations* for *every* letter in the two sequences. Consider the fact that the network is required to make different predictions when presented with the last 'S.' As stated earlier, this will only be possible if the input presented at the penultimate time step produces different internal representations in the two sequences. However, this necessary difference cannot be due to the last 'S' itself, as it is presented in both sequences. Rather, the only way for different internal representations to arise when the last 'S' is presented is when the context pool holds different patterns of activation. As the context pool holds a copy of the internal representations of the previous step, these representations must themselves be different. Recursively, we can apply the same reasoning up to the first letter. The network must therefore develop a different representation for all the letters in the sequence. Are initial different letters a sufficient condition to ensure that each letter in the sequences will be associated with different internal representations? The answer is twofold.

First, note that developing a different internal representation for each letter (including the different instances of the letter 'S') is provided *automatically* by the recurrent nature of the architecture, even without any training. Successive presentations of identical elements to a recurrent network generate different internal representations at each step because the context pool holds different patterns of activity at each step. In the above example, the first letters will generate different internal representations. On the following step, these patterns of activity will be fed back to the network, and induce different internal representations again. This process will repeat itself up to the last 'S,' and the network will therefore find itself in a state in which it is potentially able to correctly predict the last letter of the two sequences of problem [3]. Now, there is an important caveat to this observation. Another fundamental property of recurrent networks is convergence towards an attractor state when a long sequence of idential elements are presented. Even though, initially, different patterns of activation are produced on the hidden layer for each 'S' in a sequence of 'S's, eventually the network converges towards a stable state in which every new presentation of the same input produces the same pattern of activation on the hidden layer. The number of iterations required for the network to converge depends on the number of hidden units. With more degrees of freedom, it takes more iterations for the network to settle. Thus, increasing the number of hidden units provides the network with an increased *architectural* capacity of maintaining differences in its internal representations when the input elements are identical.[7]

Second, consider the way back-propagation interacts with this natural process of maintaining information about the first letter. In problem [3], the predictions in each sequence are identical up to the last letter. As similar outputs are required on each time step, the weight adjustment procedure pushes the network into developing *identical* internal representations at each time step and for the two sequences—therefore going in the opposite direction than is required. This 'homogenizing' process can strongly hinder learning, as will be illustrated below.

From the above reasoning, we can infer that optimal learning conditions exist when both contexts and predictions are different in each sequence. If the sequences share identical sequences of predictions—as in problem [3]—the process of maintaining the differences between the internal representations generated by an (initial) letter can be disrupted by back-propagation itself. The very process of learning to predict correctly the intermediate shared elements of the sequence can even cause the total error to rise sharply in some cases after an initial decrease. Indeed, the more training the network gets on these intermediate elements, the more likely it is that their internal representations will become identical, thereby completely eliminating initial slight differences that could potentially be used to disambiguate the last element. Further training can only worsen this situation.[8] Note that in this sense back-propagation in the recurrent network is not guaranteed to implement gradient descent. Presumably, the ability of the network to resist the 'homogenization' induced by the learning algorithm will depend on its representational power—the number of hidden units available for processing. With more hidden units, there is also less pressure on each unit to take on specified activation levels. Small but crucial differences in activation levels will therefore be allowed to survive at each time step, until they finally become useful at the penultimate step.

To illustrate this point, a network with fifteen hidden units was trained on the two sequences of problem [3]. The network is able to solve this problem very accurately after approximately 10,000 epochs of training on the two patterns. Learning proceeds smoothly until a very long plateau in the error is reached. This plateau corresponds to a learning phase during which the weights are adjusted so that the network can take advantage of the small differences that remain in the representations induced by the last 'S' in the two strings in order to make accurate predictions about the identity of the last letter. These slight differences are of course due to the different context generated after presentation of the first letter of the string.

To understand further the relation between network size and problem size, four different networks (with 7, 15, 30 or 120 hidden units) were trained on each of four different versions of problem [3] (with 2, 4, 6 or 12 intermediate elements). As predicted, learning was faster when the number of hidden units was larger. There was an interaction between the size of the network and the size of the problem: adding more hidden units was of little influence when the problem was small, but had a much larger impact for larger numbers of intervening 'S's. We also observed that the relation between the size of the problem and the number of epochs to reach a learning criterion was exponential for all network sizes. These results suggest that for relatively short embedded sequences of identical letters, the difficulties encountered by the simple recurrent network can be alleviated by increasing the number of hidden units. However, beyond a certain range, maintaining different representations across the embedded sequence becomes exponentially difficult (see also Allen, 1988 and Allen, 1990 for a discussion of how recurrent networks hold information across embedded sequences).

An altogether different approach to the question can also be taken. In the next section, we argue that some sequential problems may be less difficult than problem [3]. More precisely, we will show how very slight adjustments to the predictions the network is required to make in otherwise identical sequences can greatly enhance performance.

5.2. *Spanning embedded sequences*

The previous example is a limited test of the network's ability to preserve information during processing of an embedded sequence in several respects. Relevant information for making a prediction about the nature of the last letter is at a constant distance across all patterns and elements inside the embedded sequence are all identical. To evaluate the performance of the SRN on a task that is more closely related to natural language situations, we tested its ability to maintain information about long-distance dependencies on strings generated by the grammar shown in Figure 18.

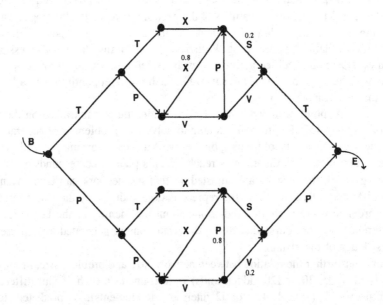

Figure 18. A complex finite-state grammar involving an embedded clause. The last letter is contingent on the first one, and the intermediate structure is shared by the two branches of the grammar. Some arcs in the asymmetrical version have different transitional probabilities in the top and bottom sub-structure as explained in the text.

If the first letter encountered in the string is a 'T,' the last letter of the string is also a 'T.' Conversely, if the first letter is a 'P,' the last letter is also a 'P.' In between these matching letters, we interposed almost the same finite state grammar that we had been using in previous experiments (Reber's) to play the role of an embedded sentence. We modified Reber's grammar by eliminating the 'S' loop and the 'T' loop in order to shorten the average length of strings.

In a first experiment we trained the network on strings generated from the finite state grammar with the same probabilities attached to corresponding arcs in the bottom and top version of Reber's grammar. This version was called the 'symmetrical grammar': contingencies inside the sub-grammar are the same independently of the first letter of the string, and all arcs had a probability of 0.5. The average length of strings was 6.5 (sd = 2.1).

After training, the performance of the network was evaluated in the following way: 20,000 strings generated from the symmetrical grammar were presented and for each string we looked at the relative activation of the predictions of 'T' and 'P' upon exit from the sub-grammar. If the Luce ratio for the prediction with the highest activation was below 0.6, the trial was treated as a 'miss' (i.e., failure to predict one or the other distinctively).[9] If the Luce ratio was greater or equal to 0.6 and the network predicted the correct alternative, a 'hit' was recorded. If the incorrect alternative was predicted, the trial was treated as an 'error.' Following training on 900,000 exemplars, performance consisted of 75 % hits, 6.3 % errors, and 18.7 % misses. Performance was best for shorter embeddings (i.e., 3 to 4 letters) and deteriorated as the length of the embedding increased (see Figure 19).

However, the fact that contingencies inside the embedded sequences are similar for both sub-grammars greatly raises the difficulty of the task and does not necessarily reflect the nature of natural language. Consider a problem of number agreement illustrated by the following two sentences:

The **dog** *that chased the cat* **is** very playful

The **dogs** *that chased the cat* **are** very playful

We would contend that expectations about concepts and words forthcoming in the embedded sentence are different for the singular and plural forms. For example, the embedded clauses require different agreement morphemes—chases vs. chase—when the clause is in the present tense, etc. Furthermore, even after the same word has been encountered in both cases (e.g., 'chased'), expectations about possible successors for that word would remain different (e.g., a single dog and a pack of dogs are likely to be chasing different things). As we have seen, if such differences in predictions do exist the network is more likely to maintain information relevant to non-local context since that information is relevant at several intermediate steps.

To illustrate this point, in a second experiment, the same network—with 15 hidden units—was trained on a variant of the grammar shown in Figure 18. In this 'asymmetrical' version, the second X arc has a 0.8 probability of being selected during training, whereas in the bottom sub-grammar, the second P arc had a probability of 0.8 of being selected. Arcs stemming from all other nodes had the same probability attached to them in both sub-grammars. The mean length of strings generated from this asymmetrical version was 5.8 letters (sd = 1.3).

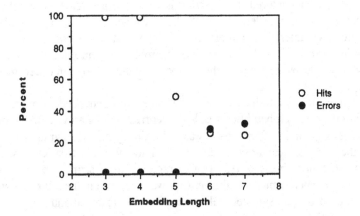

Figure 19. Percentage of hits and errors as a function of embedding length. All the cases with 7 or more letters in the embedding were grouped together.

Following training on the asymmetrical version of the grammar the network was tested with strings generated from the *symmetrical* version. Its performance level rose to 100% hits. It is important to note that performance of this network cannot be attributed to a difference in statistical properties of the test strings between the top and bottom sub-grammars—such as the difference present during training—since the testing set came from the *symmetrical* grammar. Therefore, this experiment demonstrates that the network is better able to preserve information about the predecessor of the embedded sequence across identical embeddings as long as the ensemble of *potential* pathways is differentiated during training. Furthermore, differences in potential pathways may be only statistical and, even then, rather small. We would expect even greater improvements in performance if the two sub-grammars included a set of non-overlapping sequences in addition to a set of sequences that are identical in both.

It is interesting to compare the behavior of the SRN on this embedding task with the corresponding FSA that could process the same strings. The FSA would have the structure of Figure 18. It would only be able to process the strings successfully by having two distinct copies of all the states between the initial letter in the string and the final letter. One copy is used after an initial P, the other is used after an initial T. This is inefficient since the embedded material is the same in both cases. To capture this similarity in a simple and elegant way, it is necessary to use a more powerful machine such as a recursive transition network. In this case, the embedding is treated as a subroutine which can be "called" from different places. A return from the call ensures that the grammar can correctly predict whether a T or a P will follow. This ability to handle long distance dependencies without duplication of the representation of intervening material lies at the heart of the arguments that have lead to the use of recursive formalisms to represent linguistic knowledge.

But the graded characteristics of the SRN allows the processing of embedded material as well as the material that comes after the embedding, without duplicating the representation of intervening material, and without actually making a subroutine call. The states of the SRN can be used *simultaneously* to indicate where the network is inside the embedding

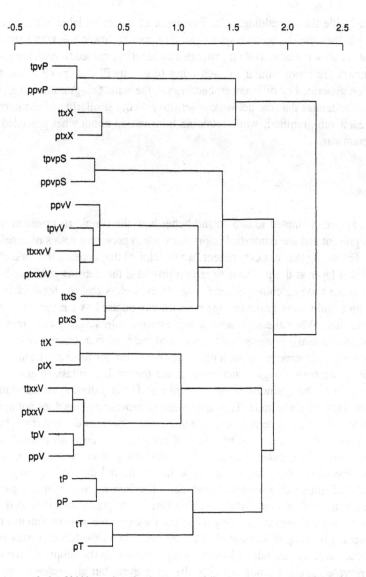

Figure 20. Cluster analysis of hidden unit activation patterns following the presentation of identical sequences in each of the two sub-grammars. Labels starting with the letter 't' come from the top sub-grammars, labels starting with the letter 'p' come from the bottom sub-grammar.

and to indicate the history of processing prior to the embedding. The identity of the initial letter simply *shades* the representation of states inside the embedding, so that corresponding nodes have similar representations, and are processed using overlapping portions of the knowledge encoded in the connection weights. Yet the shading that the initial letter provides allows the network to carry information about the early part of the string through the embedding, thereby allowing the network to exploit long-distance dependencies. This property of the internal representations used by the SRN is illustrated in Figure 20. We recorded some patterns of activation over the hidden units following the presentation of each letter inside the embeddings. The first letter of the string label in the figure (t or p) indicates whether the string corresponds to the upper or lower sub-grammar. The figure shows that the patterns of activation generated by identical embeddings in the two different sub-grammars are more similar to each other (e.g., 'tpvP' and 'ppvP') than to patterns of activation generated by different embeddings in the same sub-grammar (e.g., 'tpvP' and 'tpV'). This indicates that the network is sensitive to the similarlity of the corresponding nodes in each sub-grammar, while retaining information about what preceded entry into the sub-grammar.

6. Discussion

In this study, we attempted to understand better how the simple recurrent network could learn to represent and use contextual information when presented with structured sequences of inputs. Following the first experiment, we concluded that copying the state of activation on the hidden layer at the previous time step provided the network with the basic equipment of a finite state machine. When the set of exemplars that the network is trained on comes from a finite state grammar, the network can be used as a recognizer with respect to that grammar. When the representational resources are severely constrained, internal representations actually converge on the nodes of the grammar. Interestingly, though, this representational convergence is not a necessary condition for functional convergence: networks with more than enough structure to handle the prediction task sometimes represent the same node of the grammar using two quite different patterns, corresponding to different paths into the same node. This divergence of representations does not upset the network's ability to serve as a recognizer for well-formed sequences derived from the grammar.

We also showed that the mere presence of recurrent connections pushed the network to develop hidden layer patterns that capture information about sequences of inputs, even in the absence of training. The second experiment showed that back-propagation can be used to take advantage of this natural tendency when information about the path traversed is relevant to the task at hand. This was illustrated with predictions that were specific to particular subsequences in the training set or that took into account constraints on the length of sequences. Encoding of sequential structure depends on the fact that back-propagation causes hidden layers to encode task-relevant information. In the simple recurrent network, internal representations encode not only the prior event but also relevant aspects of the representation that was constructed in predicting the prior event from its predecessor. When fed back as input, these representations provide information that allows the network to maintain prediction-relevant features of an entire sequence. We illustrated this with cluster analyses of the hidden layer patterns.

Our description of the stages of learning suggested that the network initially learns to distinguish between events independently of the temporal context (e.g., simply distinguish between different letters). The information contained in the context layer is ignored at this point. At the end of this stage, each event is associated to a specific pattern on the hidden layer that identifies it for the following event. In the next phase, thanks to this new information, different occurrences of the same event (e.g., two occurrences of the same letter) are distinguished on the basis of immediately preceding events—the simplest form of a time tag. This stage corresponds to the recognition of the different 'arcs' in the particular finite state grammar used in the experiments. Finally, as the representation of each event progressively acquires a time tag, sub-sequences of events come to yield characteristic hidden layer patterns that can form the basis of further discriminations (e.g., between an 'early' and 'late' 'T$_2$' in the Reber grammar). In this manner, and under appropriate conditions, the hidden unit patterns achieve an encoding of the entire sequence of events presented.

We do not mean to suggest that simple recurrent networks can learn to recognize *any* finite state language. Indeed, we were able to predict two conditions under which performance of the simple recurrent network will deteriorate: (1) when different sequences may contain identical embedded sequences involving *exactly* the same predictions; and (2) when the number of hidden units is restricted and cannot support redundant representations of similar predictions, so that identical predictions following different events tend to be associated with very similar hidden unit patterns, thereby erasing information about the initial path. We also noted that when recurrent connections are added to a three-layer feed-forward network, back-propagation is no longer guaranteed to perform gradient descent in the error space. Additional training, by improving performance on shared components of otherwise differing sequences, can eliminate information necessary to 'span' an embedded sequence and result in a sudden rise in the total error. It follows from these limitations that the simple recurrent network could not be expected to learn sequences with a moderately complex recursive structure—such as context free grammars—if contingencies inside the embedded structures do not depend on relevant information preceding the embeddings.

What is the relevance of this work with regard to language processing? The ability to exploit long-distance dependencies is an inherent aspect of human language processing capabilities, and it lies at the heart of the general belief that a recursive computational machine is necessary for processing natural language. The experiments we have done with SRNs suggest another possibility; it may be that long-distance dependencies can be processed by machines that are simpler than fully recursive machines, as long as they make use of *graded* state information. This is particularly true if the probability structure of the grammar defining the material to be learned reflects—even very slightly—the information that needs to be maintained. As we noted previously, natural linguistic stimuli may show this property. Of course, true natural language is far more complex than the simple strings that can be generated by the machine shown in Figure 24, so we cannot claim to have shown that graded state machines will be able to process all aspects of natural language. However, our experiments indicate already that they are more powerful in interesting ways than traditional finite state automata (see also the work of Allen and Riecksen, 1989; Elman, 1990; and Pollack, in press). Certainly, the SRN should be seen as a new entry into the taxonomy of computational machines. Whether the SRN—or rather some other instance of the broader class of graded state machines of which the SRN is one of the simplest—will ultimately turn out to prove sufficient for natural language processing remains to be explored by further research.

Acknowledgments

We gratefully acknowledge the constructive comments of Jordan Pollack and an anonymous reviewer on an earlier draft of this paper. David Servan-Schreiber was supported by an

NIMH Individual Fellow Award MH-09696-01. Axel Cleeremans was supported by a grant from the National Fund for Scientific Research (Belgium). James L. McClelland was supported by an NIMH Research Scientist Career Development Award MH-00385. Support for computational resources was provided by NSF(BNS-86-09729) and ONR(N00014-86-G-0146). Portions of this paper have previously appeared in Servan-Schreiber, Cleeremans & McClelland (1989), and in Cleeremans, Servan-Schreiber & McClelland (1989).

Notes

1. Slightly modified versions of the BP program from McClelland and Rumelhart (1988) were used for this and all subsequent simulations reported in this paper. The weights in the network were initially set to random values between −0.5 and +0.5. Values of learning rate and momentum (*eta* and *alpha* in Rumelhart el al. (1986)) were chosen sufficiently small to avoid large oscillations and were generally in the range of 0.01 to 0.02 for learning rate and 0.5 to 0.9 for momentum.

2. For any single output unit, given that targets are binary, and assuming a fixed input pattern for all training exemplars, the error can be expressed as:

$$p(1 - a)^2 + (1 - p)a^2$$

where p is the probability that the unit should be on, and a is the activation of the unit. The first term applies when the target is 1, the second when the target is 0. Back-propagation tends to minimize the derivative of this expression, which is simply $2a - 2p$. The minimum is attained when $a = p$, i.e., when the activation of the unit is equal to its probability of being on in the training set (Rumelhart, personal communication to McClelland, Spring 1989).

3. Cluster analysis is a method that finds the optimal partition of a set of vectors according to some measure of similarity (here, the euclidean distance). On the graphical representation of the obtained clusters, the contrast between two groups is indicated by the length of the horizontal links. The length of vertical links is not meaningful.

4. This fact may seem surprising at first, since the learning algorithm does not apply pressure on the weights to generate different representations for different paths to the same node. Preserving that kind of information about the path does not contribute in itself to reducing error in the prediction task. We must therefore conclude that this differentiation is a direct consequence of the recurrent nature of the architecture rather than a consequence of back-propagation. Indeed, in Servan-Schreiber, Cleeremans and McClelland (1988), we showed that some amount of information about the path is encoded in the hidden layer patterns when a succession of letters is presented, even in the absence of any training.

5. In fact, length constraints are treated exactly as atypical cases since there is no representation of the length of the string as such.

6. *For each statistical order*, the analysis consisted of three steps: First, we estimated the conditional probabilities of observing each letter after each possible path through the grammar (e.g., the probabilities of observing each of the seven letters given the sequence 'TSS'). Second, we computed the probabilities that each of the above paths leads to each node of the grammar (e.g., the probabilities that the path "TSS" finishes at node #1, node #2, etc.). Third, we obtained the average conditional probabilities (ACP) of observing each letter at each node of the grammar by summing the products of the terms obtained in steps #1 and #2 over the set of possible paths. Finally, all the ACPs that corresponded to letters that could *not* appear at a particular node (e.g., a 'V' at node #0, etc.) were eliminated from the analysis. Thus, for each statistical order, we obtained a set of 11 ACPs (one for each occurrence of the five letters, and one for 'E,' which can only appear at node #6. 'B' is never predicted).

7. For example, with three hidden units, the network converges to a stable state after an average of three iterations when presented with identical inputs (with a precision of two decimal points for each unit). A network with 15 hidden units converges after an average of 8 iterations. These results were obtained with random weights in the range [−0.5, +0.5].

8. Generally, small values for the learning rate and momentum, as well as many hidden units, help to minimize this problem.
9. The Luce ratio is the ratio of the highest activation on the output layer to the sum of all activations on that layer. This measure is commonly applied in psychology to model the strength of a response tendency among a finite set of alternatives (Luce, 1963). In this simulation, a Luce ratio of 0.5 often corresponded to a situation where 'T' and 'P' were equally activated and all other alternatives were set to zero.

References

Allen, R.B. (1988). Sequential connectionist networks for answering simple questions about a microworld. *Proceedings of the Tenth Annual Conference of the Cognitive Science Society.*

Allen, R.B., & Riecksen, M.E. (1989). Reference in connectionist language users. In R. Pfeifer, Z. Schreter, F. Fogelman-Soulié, & L. Steels (Eds.), *Connectionism in perspective.* North Holland: Amsterdam.

Allen, R.B. (1990). *Connectionist language users* (TR-AR-90-402). Morristown, NJ: Bell Communications Research.

Cleeremans A., Servan-Schreiber D., & McClelland J.L. (1989). Finite state automata and simple recurrent networks. *Neural Computation, 1,* 372–381.

Cottrell, G.W. (1985). Connectionist parsing. *Proceedings of the Seventh Annual Conference of the Cognitive Science Society.* Hillsdale, NJ: Erlbaum.

Elman, J.L. (1990). Finding structure in time. *Cognitive Science,* 14, 179–211.

Elman, J.L. (1990). Representation and structure in connectionist models. In Gerry T.M. Altmann (Ed.), *Cognitive models of speech processing: Psycholinguistic and computational perspectives.* Cambridge, MA: MIT Press.

Fanty, M. (1985). *Context-free parsing in connectionist networks* (TR174). Rochester, NY: University of Rochester, Computer Science Department.

Hanson, S. & Kegl, J. (1987). PARSNIP: A connectionist network that learns natural language from exposure to natural language sentences. *Proceedings of the Ninth Annual Conference of the Cognitive Science Society.* Hillsdale, NJ: Erlbaum.

Hinton, G., McClelland, J.L., & Rumelhart, D.E. (1986). Distributed representations. In D.E. Rumelhart and J.L. McClelland (Eds.), *Parallel distributed processing, I: Foundations.* Cambridge, MA: MIT Press.

Jordan, M.I. (1986). Attractor dynamics and parallelism in a connectionist sequential machine. *Proceedings of the Eighth Annual Conference of the Cognitive Science Society.* Hillsdale, NJ: Erlbaum.

Luce, R.D. (1963). Detection and recognition. In R.D. Luce, R.R. Bush and E. Galanter (Eds.), *Handbook of mathematical psychology (Vol. 1).* New York: Wiley.

McClelland, J.L., & Rumelhart, D.E. (1988). *Explorations in parallel distributed processing: A handbook of models, programs and exercises.* Cambridge, MA: MIT Press.

Pollack, J. (in press). Recursive distributed representations. *Artificial Intelligence.*

Reber, A.S. (1976). Implicit learning of synthetic languages: The role of the instructional set. *Journal of Experimental Psychology: Human Learning and Memory, 2,* 88–94.

Rumelhart, D.E., & McClelland, J.L. (1986). *Parallel distributed processing, I: Foundations.* Cambridge, MA: MIT Press.

Rumelhart, D.E., Hinton, G., & Williams, R.J. (1986). Learning internal representations by error propagation. In D.E. Rumelhart and J.L. McClelland (Eds.), *Parallel distributed processing, I: Foundations.* Cambridge, MA: MIT Press.

Sejnowski, T.J., & Rosenberg, C. (1987). Parallel networks that learn to pronounce english text. *Complex Systems, 1,* 145–168.

Servan-Schreiber D., Cleeremans A., & McClelland J.L. (1988). *Encoding sequential structure in simple recurrent networks* (Technical Report CMU-CS-183). Pittsburgh, PA: Carnegie Mellon University, School of Computer Science.

Servan-Schreiber D., Cleeremans A., & McClelland J.L. (1989). Learning sequential structure in simple recurrent networks. In D.S. Touretzky (Ed.), *Advances in neural information processing systems 1.* San Mateo, CA: Morgan Kaufmann. [Collected papers of the IEEE Conference on Neural Information Processing Systems— Natural and Synthetic, Denver, Nov. 28–Dec. 1, 1988].

St. John, M., & McClelland, J.L. (in press). Learning and applying contextual constraints in sentence comprehension. *Artificial Intelligence.*

Machine Learning, 7, 195–225 (1991)

Distributed Representations, Simple Recurrent Networks, and Grammatical Structure

JEFFREY L. ELMAN   (ELMAN@CRL.UCSD.EDU)
Departments of Cognitive Science and Linguistics, University of California, San Diego

Abstract. In this paper three problems for a connectionist account of language are considered:

1. What is the nature of linguistic representations?
2. How can complex structural relationships such as constituent structure be represented?
3. How can the apparently open-ended nature of language be accommodated by a fixed-resource system?

Using a prediction task, a simple recurrent network (SRN) is trained on multiclausal sentences which contain multiply-embedded relative clauses. Principal component analysis of the hidden unit activation patterns reveals that the network solves the task by developing complex distributed representations which encode the relevant grammatical relations and hierarchical constituent structure. Differences between the SRN state representations and the more traditional pushdown store are discussed in the final section.

Keywords. Distributed representations, simple recurrent networks, grammatical structure

1. Introduction

In recent years there has been considerable progress in developing connectionist models of language. This work has demonstrated the ability of network models to account for a variety of phenomena in phonology (e.g., Gasser & Lee, 1990; Hare, 1990; Touretzky, 1989; Touretzky & Wheeler, 1989), morphology (e.g., Hare, Corina, & Cottrell, 1989; MacWhinney et al. 1989; Plunkett & Marchman, 1989; Rumelhart & McClelland, 1986b; Ryder, 1989), spoken word recognition (McClelland & Elman, 1986), written word recognition (Rumelhart & McClelland, 1986; Seidenberg & McClelland, 1989), speech production (Dell, 1986; Stemberger, 1985), and role assignment (Kawamoto & McClelland, 1986; Miikkulainen & Dyer, 1989a; St. John & McClelland, 1989). It is clear that connectionist networks have many properties which make them attractive for language processing.

At the same time, there remain significant shortcomings to current work. This is hardly surprising: natural language is a very difficult domain. It poses difficult challenges for any paradigm. These challenges should be seen in a positive light. They test the power of the framework and can also motivate the development of new connectionist approaches.

In this paper I would like to focus on what I see as three of the principal challenges to a successful connectionist account of language. They are:

1. *What is the nature of the linguistic representations?*
2. *How can complex structural relationships such as constituency be represented?*
3. *How can the apparently open-ended nature of language be accommodated by a fixed-resource system?*

Interestingly, these problems are closely intertwined, and all have to do with representation.

One approach which addresses the first two problems is to use localist representations. In localist networks, nodes are assigned discrete interpretations. In such models (e.g., Kawamoto & McClelland, 1986; St. John & McClelland, 1988) nodes may represent grammatical roles (e.g., agent, theme, modifier) or relations (e.g., subject, daughter-of). These may be then bound to other nodes which represent the word-tokens which instantiate them either by spatial assignment (Kawamoto & McClelland, 1986; Miikkulainen & Dyer, 1989b), concurrent activation (St. John & McClelland, 1989), or various other techniques (e.g., Solensky, in press).

Although the localist approach has many attractions, it has a number of important drawbacks as well.

First, the localist dictum, "one node/one concept," when taken together with the fact that networks typically have fixed resources, seems to be at variance with the open-ended nature of language. If nodes are pre-allocated to defined roles such as subject or agent, then in order to process sentences with multiple subjects or agents (as is the case with complex sentences) there must be the appropriate number and type of nodes. But how is one to know just which types will be needed, or how many to provide? The situation becomes even more troublesome if one is interested in discourse phenomena. Generative theories of language (Chomsky, 1965), have made much of the unbounded generativity of natural language; it has been pointed out (Rumelhart & McClelland, 1986a) that in reality, language productions in practice are in fact of finite length and number. Still, even if one accepts these the practical limitations, it is noteworthy that they are soft (or context-sensitive), rather than hard (or absolute) in the way that the localist approach would predict. (For instance, consider the difficulty of understanding "the cat the dog the mouse saw chased ran away" compared with, "the planet the astronomer the university hired saw exploded." Clearly, semantic and pragmatic considerations can facilitate parsing structures which are otherwise hard to process (see also Labov, 1973; Reich & Dell, 1977; Schlesinger, 1968; Stolz, 1967, for experimental demonstrations of this point). Thus, although one might anticipate the most commonly occurring structural relations one would like the limits on processing to be soft rather than hard, in the way the localist approach would be.

A second shortcoming to the use of localist representations is that they often underestimate the actual richness of linguistic structure. Even the basic notion "word," which one might assume to be a straightforward linguistic primitive, turns out to be more difficult to define than one might have thought. There are dramatic differences in terms of what counts as a word across languages; and even within English, there are morphological and syntactic processes which yield entities which are word-like in some but not all respects (e.g., apple pie, man-in-the-street, man for all seasons). In fact, much of linguistic theory is today concerned with the nature and role of representation, with less focus on the nature of operations.

Thus, while the localist approach has certain positive aspects, it has definite shortcomings as well. It provides no good solution to the problem of how to account for the open-ended nature of language, and the commitment to discrete and well-defined representations may make it difficult to capture the richness and high dimensionality required for language representations.

Another major approach involves the use of distributed representations (Hinton, 1988; Hinton, McClelland, & Rumelhart, 1986; van Gelder, in press), together with a learning algorithm, in order to infer the linguistic representations. Models which have used the localist approach have typically made an *a priori* commitment to linguistic representations (such as agent, patient, etc.); networks are then explicitly trained to identify these representations in the input by activating nodes which correspond to them. This presupposes that the target representations are theoretically valid; it also begs the question of where (in the real world) the corresponding teaching information might come from. In the alternative approach, tasks must be devised in which the abstract linguistic representations do not play an explicit role. The model's inputs and output targets are limited to variables which are directly observable in the environment. This is a more naturalistic approach in the sense that the model learns to use surface linguistic forms for communicative purposes rather than to do linguistic analysis. Whatever linguistic analysis is done (and whatever representations are developed) is internal to the network and is in the service of a task. The value of this approach is that it need not depend on preexisting preconceptions about what the abstract linguistic representations are. Instead, the connectionist model can be seen as a mechanism for gaining new theoretical insight. Thus, this approach offers a potentially more satisfying answer to the first question, What are the nature of linguistic representations?

There is a second advantage to this approach. Because the abstract representations are formed at the hidden layer, they also tend to be distributed across the high-dimensional (and continuous) space which is described by analog hidden unit activation vectors. This means there is a larger and much finer-grained representational space to work with than is usually possible with localist representations. This space is not infinite, but for practical purposes it may be very, very large. And so this approach may also provide a better response to the third question, How can the apparently open-ended nature of language be accommodated by a fixed-resource system?

But all is not rosy. We are still left with the second question: How to represent complex structural relationships such as constituency. Distributed representations are far more complex and difficult to understand than localist representations. There has been some tendency to feel that their murkiness is intractable and that "distributed" entails "unanalyzable." Although, in fact, there exist various techniques for analyzing distributed representations (including cluster analysis, Elman, 1990; Hinton, 1988; Sejnowski & Rosenberg, 1987; Servan-Schreiber, Cleeremans, & McClelland, in press; direct inpsection, Pollack, 1988; principal component and phase state analysis, Elman, 1989; and contribution analysis, Sanger, 1989), the results of such studies have been limited. These analyses have demonstrated that distributed representations may possess internal structure which can encode relationships such as kinship (Hinton, 1987) or lexical category structure (Elman, 1990). But such relationships are static. Thus, for instance, in Elman (1990) a network was trained to predict the order of words in sentences. The network learned to represent words by categorizing them as nouns or verbs, with further subcategorization of nouns as animate/inanimate, human/non-human, etc. These representations were developed by the network and were not explicitly taught.

While lexical categories are surely important for language processing, it is easy to think of other sorts of categorization which seem to have a different nature. Consider the following sentences.

(1a) The boy broke the window.
(1b) The rock broke the window.
(1c) The window broke.

The underlined words in all the sentences are nouns, and their representations should reflect this. Nounhood is a category property which belongs inalienably to these words, and is true of them regardless of where they appear (as nouns; derivational processes may result in nouns being used as verbs, and *vice versa*). At a different level of description, the underlined words are also similar in that they are categorizable as the subjects of their sentences. This property, however, is context-dependent. The word "window" is a subject only in sentence (1c). In the other two sentences it is an object. At still another level of description, the three underlined words differ. In (1a) the subject is also the agent of the event; in (1b) the subject is the instrument; and in (1c) the subject is the patient (or theme) of the sentence. This too is a context-dependent property.

These examples are simple demonstrations of the effect of grammatical structure; that is, structure which is manifest at the level of utterance. In addition to their context-free categorization, words inherit properties by virtue of their linguistic environment. Although distributed representations seem potentially able to respond to the first and last of the problems posed at the outset, it is not clear how they address the question, How can complex structural relationships such as constituency be represented? As Fodor & Pylyshyn (1988) have phrased it,

> You need two degrees of freedom to specify the thoughts that an intentional system is entertaining at a time; one parameter (active vs inactive) picks out the nodes that express concepts that the system has in mind; the other (in construction vs not) determines how the concepts that the system has in mind are distributed in the propositions that it entertains. (pp. 25–26)

At this point, it is worth reminding ourselves of the ways in which complex structural relationships are dealt with in symbolic systems. Context-free properties are typically represented with abstract symbols such as S, NP, V, etc. Context-sensitive properties are dealt with in various ways. Some theories (e.g., Generalized Phrase Structure Grammar) designate the context in an explicit manner through so-called "slash-categories." Other approaches use additional category labels (e.g., Cognitive Grammar, Relational Grammar, Government and Binding) to designate elements as subject, theme, argument, trajectory, path, etc. In addition, theories may make use of trees, bracketing, co-indexing, spatial organization, tiers, arcs, circles, and diacritics in order to convey more complex relationships and mappings. Processing or implementation versions exist for some of these theories; nearly all require a working buffer or stack in order to account for the apparently recursive nature of utterances. All in all, a rather formidable armamentarium is required.

Returning to the three questions posed at the outset, although distributed representations have characteristics which plausibly may address the need for representational richness, flexibility, and may provide soft (rather than hard) limits on processing; but we now must ask whether such an approach can capture structural relationships of the sort required for language. That is the question which motivated the work to be reported here.

There is preliminary evidence which is encouraging in this regard. Hinton (1988) has described a scheme which involve "reduced descriptions" of complex structures, and which represent part-whole hierarchies. Pollack (1988, in press) has developed a training regimen called Recursive Auto-Associative Memory (RAAM) which appears to have compositional properties and which supports structure-sensitive operations (see also Chalmers, 1989). As discussed earlier, Elman's (1990) use of Simple Recurrent Networks (SRN; Servan-Schreiber, Cleeremans, & McClelland, in press) provides yet another approach for encoding structural relationships in a distributed form.

The work described here extends this latter approach. An SRN was taught a task involving stimuli in which there were underlying hierarchical (and recursive) relationships. This structure was abstract in the sense that it was implicit in the stimuli, and the goal was to see if the network could (a) infer this abstract structure; and (b) represent the compositional relationships in such a manner as to support structure-sensitive operations.

The remainder of this paper is organizaed as follows. First, the network architecture will be briefly introduced. Second, the stimulus set and tast will be presented, and the properties of the task which make it particularly relevant for the question at hand will be described. Next, the results of the simulation will be presented. In the final discussion, differences and similarities between this approach and more traditional symbolic approaches to language processing will be discussed.

2. Network architecture

Time is an important element in language, and so the question of how to represent serially ordered inputs is crucial. Various proposals have been advanced (for reviews, see Elman, 1990; Mozer, 1988). The approach taken here involves treating the network as a simple dynamical system in which previous states are made available as an additional input (Jordan, 1986). In Jordan's work, the network state at any point in time was a function of the input on the current time step, plus the state of the output units on the previous time step. In the work here, the network's state depends on current input, plus its own internal state (represented by the hidden units) on the previous cycle. Because the hidden units are not taught to assume specific values, this means that they can develop representations, in the course of learning a task, which encode the temporal structure of the task. In other words, the hidden units learn to become a kind of memory which is very task-specific.

The type of network used in the current work is shown in Figure 1. This network has the typical connections from **input units** to **hidden units**, and from hidden units to **output units**. (Additional hidden layers between input and main hidden, and between main hidden and output, may be used to serve as transducers which compress the input and output vectors.) There are an additional set of units, called **context units**, which provide for limited recurrence (and so this may be called a **simple recurrent network**). These context units are activated on a one-for-one basis by the hidden units, with a fixed weight of 1.0, and have linear activation functions.

The result is that at each time cycle the hidden unit activations are copied into the context units; on the next time cycle, the context combines with the new input to activate the hidden units. The hidden units therefore take on the job of mapping new inputs and prior

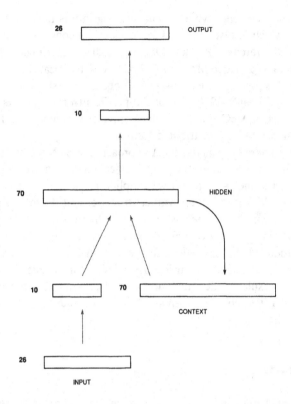

Figure 1. Network architecture. Hidden unit activations are copied along fixed weights (of 1.0) into linear Context units on a one-to-one basis; on the next time step the Context units feed into Hidden units on a distributed basis. Additional hidden units between input and main hidden layer, and between main hidden layer and output, provide compress basis vectors into more compact form.

states to the output. Because they themselves constitute the prior state, they must develop representations which facilitate this input/output mapping. The simple recurrent network has been studied in a number of tasks (Elman, 1990; Gasser, 1989; Hare, Corina, & Cottrell, 1988; Servan-Schreiber, Cleeremans, & McClelland, in press).

3. Task and stimuli

3.1. The prediction task

In Elman (1990) a network similar to that in Figure 1 was trained to predict the order of words in simple (2- and 3-word) sentences. At each point in time, a word was presented to the network. The network's target output was simply the next word in sequence. The lexical items (inputs and outputs) were represented in a localist form using basis vectors; i.e., each word was randomly assigned a vector in which a single bit was turned on. Lexical items were thus orthogonal to one another, and the form of each item did not encode any information about the item's category membership. The prediction was made on the basis of the current input word, together with the prior hidden unit state (saved in the context units).

This task was chosen for several reasons. First, the task meets the desideratum that the inputs and target outputs be limited to observables in the environment. The network's inputs and outputs are immediately available and require minimal a priori theoretical analysis (lexical items are orthogonal and arbitrarily assigned). The role of an external teacher is minimized, since the target outputs are supplied by the environment at the next moment in time. The task involves what might be called "self-supervised learning."

Second, although language processing obviously involves a great deal more than prediction, prediction does seem to play a role in processing. Listeners can indeed predict (Grosjean, 1980), and sequences of words which violate expectations—i.e., which are unpredictable—result in distinctive electrical activity in the brain (Kutas, 1988; Kutas & Hillyard, 1980; Tanenhaus et al., in press).

Third, if we accept that prediction or anticipation plays a role in language learning, then this provides a partial solution to what has been called Baker's paradox (Baker, 1979; Pinker, 1989). The paradox is that children apparently do not receive (or ignore, when they do) negative evidence in the process of language learning. Given their frequent tendency initially to over-generalize from positive data, it is not clear how children are able to retract the faulty over-generalizations (Gold, 1967). However, if we suppose that children make covert predictions about the speech they will hear from others, then failed predictions constitute an indirect source of negative evidence which could be used to refine and retract the scope of generalization.

Fourth, the task requires that the network discover the regularities which underlie the temporal order of the words in the sentences. In the simulation reported in Elman (1990) these regularities resulted in the network's constructing internal representations of inputs which marked words for form class (noun/verb) as well as lexico-semantic characteristics (animate/inanimate, human/animal, large/small, etc.)

The results of that simulation, however, bore more on the representation of lexical category structure, and the relevance to grammatical structure is unclear. Only monoclausal sentences were used, all shared the same basic structure. Thus the question remains open whether the internal representations that can be learned in such an architecture are able to encode the hierarchical relationships which are necessary to mark constituent structure.

3.2. Stimuli

The stimuli in this simulation were sequences of words which were formed into sentences. In addition to monoclausal sentences, there were a large number of complex multi-clausal sentences.

Sentences were formed from a lexicon of 23 items. These included 8 nouns, 12 verbs, ther relative pronoun **who**, and an end-of-sentence indicator (a period). Each item was represented by a randomly assigned 26-bit vector in which a single bit was set to 1 (3 bits were reserved for another purpose). A phrase structure grammar, shown in Table 1, was used to generate sentences. The resulting sentences possessed certain important properties. These include the following.

Table 1.

S → NP VP "."
NP → PropN | N | N RC
VP → V (NP)
RC → *who* NP VP | *who* VP (NP)
N → *boy* | *girl* | *cat* | *dog* | *boys* | *girls* | *cats* | *dogs*
PropN → *John* | *Mary*
V → *chase* | *feed* | *see* | *hear* | *walk* | *live* | *chases* | *feeds* | *sees* | *hears* | *walks* | *lives*

Additional restrictions:
  • number agreement between N & V within clause, and (where appropriate) between
    head N & subordinate V
  • verb arguments:
      chase, feed → require a direct object
      see, hear → optionally allow a direct object
      walk, live → preclude a direct object
      (observed also for head/verb relations in relative clauses

---

3.2.1. Agreement

Subject nouns agree with their verbs. Thus, for example, (2a) is grammatical but not (2b). (The training corpus consisted of positive examples only; starred examples below did not actually occur).

(2a) John feeds dogs.
(2b) *Boys sees Mary.

Words are not marked for number (singular/plural), form class (verb/noun, etc.), or grammatical role (subject/object, etc.). The network must learn first that there are items which function as what we would call nouns, verbs, etc.; then it must learn which items are examples of singular and plural; and then it must learn which nouns are subjects and which are objects (since agreement only holds between subject nouns and their verbs).

3.2.2. Verb argument structure

Verbs fall into three classes: those that require direct objects, those that permit an optional direct object, and those that preclude direct objects. As a result, sentences (3a–d) are grammatical, whereas sentences (3e, 3f) are ungrammatical.

(3a) Girls feed dogs. (*D.O. required*)
(3b) Girls see boys. (*D.O. optional*)
(3c) Girls see. (*D.O. optional*)
(3d) Girls live. (*D.O. precluded*)
(3e) *Girls feed.
(3f) *Girls live dogs.

Because all words are represented with orthogonal vectors, the type of verb is not overtly marked in the input and so the class membership needs to be inferred at the same time as the cooccurrence facts are learned.

3.2.3. Interactions with relative clauses

The agreement and the verb argument facts become more complicated in relative clauses. Although direct objects normally follow the verb in simple sentences, some relative clauses have the subordinate clause direct object as the head of the clause. In these cases, the network must recognize that there is a gap following the subordinate clause verb (because the direct object role has already been filled. Thus, the normal pattern in simple sentences (3a–d) appears also in (4a), but contrasts with (4b),

(4a) Dog _{who chases cat} sees girl.

(4b) Dog _{who cat chases} sees girl.

On the other hand, sentence (4c), which seems to conform to the pattern established in (3) and (4a), is ungrammatical.

(4c) *Dog _{who cat chases dog} sees girl.

Similar complicatons arise for the agreements facts. In simple declarative sentences agreement involves $N1 - V1$. In complex sentences, such as (5a), that regularity is violated, and any straightforward attempt to generalize it to sentences with multiple clauses would lead to the ungrammatical (5b).

(5a) Dog _{who boys feed} sees girl.

(5b) *Dog _{who boys feeds} see girl.

3.2.4. Recursion

The grammar permits recursion through the presence of relative clause (which expand to noun phrases which may introduce yet other relative clauses, etc.). This leads to sentences such as (6) in which the grammatical phenomena noted in (a–c) may be extended over a considerable distance.

(6) Boys _{who girls} _{who dogs chase} see hear.

3.2.5. Viable sentences

One of the literals inserted by the grammar is ",", which occurs at the end of sentences. This end-of-sentence marker can potentially occur anywhere in a string where a grammatical sentence might be terminated. Thus in sentence (7), the carets indicate positions where a "." might legally occur.

(7) Boys see ∧ dogs ∧ who see ∧ girls ∧ who hear ∧ .

The data in (4–7) are examples of the sorts of phenomena which linguists argue cannot be accounted for without abstract representations. More precisely, it has been claimed that such abstract representations offer a more perspicacious account of grammatical phenomena than one which, for example, simply lists the surface strings (Chomsky, 1957).

The training data were generated from the grammar summarized in Table 1. At any given point during training, the training set consisted of 10,000 sentences which were presented to the network 5 times. (As before, sentences were concatenated so that the input stream proceeded smoothly without breaks between sentences.) However, the composition of these sentences varied over time. The following training regimen was used in order to provide for incremental training. The network was trained on 5 passes through each of the following 4 corpora.

Phase 1: The first training set consisted exclusively of simple sentences. This was accomplished by eliminating all relative clauses. The result was a corpus of 34,605 words forming 10,000 sentences (each sentence includes the terminal " . ").

Phase 2: The network was then exposed to a second corpus of 10,000 sentences which consisted of 25% complex sentences and 75% simple sentences (complex sentences were obtained by permitting relative clauses). Mean sentence length was 3.92 (minimum: 3 words, maximum: 13 words).

Phase 3: The third corpus increased the percentage of complex sentences to 50%, with mean sentence length of 4.38 (minimum: 3 words, maximum: 13 words).

Phase 4: The fourth consisted of 10,000 sentences, 75% complex, 25% simple. Mean sentence length was 6.02 (minimum: 3 words, maximum: 16 words).

This staged learning strategy was developed in response to results of earlier pilot work. In this work, it was found that the network was unable to learn the task when given the full range of complex data from the beginning of training. However, when the network was permitted to focus on the simpler data first, it was able to learn the task quickly and then move on successfully to more complex patterns. The important aspect to this was that the earlier training constrained later learning in a useful way; the early training forced the network to focus on canonical versions of the problems which apparently created a good basis for then solving the more difficult forms of the same problems.

4. Results

At the conclusion of the fourth phase of training, the weights were frozen at their final values and network performance was tested on a novel set of data, generated in the same way as the last training corpus. Becaus the task is non-deterministic, the network will (unless it memorizes the sequence) always produce errors. The optimal strategy in this case will be to activate the output units (i.e., predict potential next words) to some extent proportional to their statistical likelihood of occurrence. Therefore, rather than assessing the network's global performance by looking at root mean squared error, we should ask how closely the network approximated these probabilities. The technique described in Elman (in press) was used to accomplish this. Context-dependent likelihood vectors were generated for each word in every sentence; these vectors represented the empirically derived probabilities of occurrence for all possible predictions, given the sentence context up to that point. The network's actual outputs were then compared against these likelihood vectors, and this error was used to measure performance. The error was quite low: 0.177 (initial error: 12.45; minimal error through equal activation of all units would be 1.92). This error can also be normalized by computing the mean cosine of the angle between the vectors, which is 0.852 (sd: 0.259). Both measures indicate that the network achieved a high level of performance in prediction.

These gross measures of performance, however, do not tell us how well the network has done in each of the specific problem areas posed by the task. Let us look at each area in turn.

4.1. Agreement in simple sentences

Agreement in simple sentences is shown in Figures 2a and 2b.

The network's predictions following the word **boy** are that either a singular verb will follow (words in all three singular verb categories are activated, since it has no basis for predicting the type of verb), or else that the next word may be the relative pronoun **who**. Conversely, when the input is the word **boys**, the expectation is that a verb in the plural will follow, or else the relative pronoun. (Similar expectations hold for the other nouns in the lexicon. In this and the results that follow, the performance of the sentences which are shown is representative over other sentences with similar structure.)

4.2. Verb argument structure in simple sentences

Figure 3 shows network predictions following an initial noun and then a verb from each of the three different verb types.

When the verb is **lives**, the network's expectation is that the following item will be " . " (which is in fact the only successor permitted by the grammar in this context). The verb **sees**, on the other hand, may either be followed by a " . ", or optionally by a direct object (which may be a singular or plural noun, or proper noun). Finally, the verb **chases** requires a direct object, and the network learns to expect a noun following this and other verbs in the same class.

4.3. Interactions with relative clauses

The examples so far have all involved simple sentences. The agreement and verb argument facts are more complicated in complex sentences. Figure 4 shows the network predictions for each word in the sentence **boys who mary chases feed cats**. If the network were generalizing the pattern for agreement found in the simple sentences, we might expect the network to predict a singular verb following ...**mary chases**... (insofar as it predicts a verb in this position at all; conversely, it might be confused by the pattern *N1 N2 V1*).

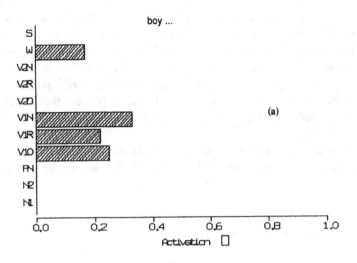

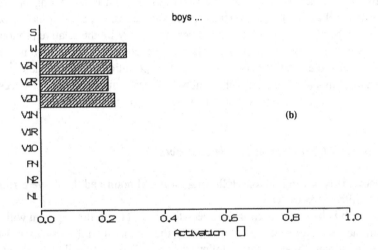

Figure 2. (a) Graph of network predictions following presentation of the word **boy**. Predictions are shown as activations for words grouped by category. S stands for end-of-sentence (" . "); W stands for who; N and V represent nouns and verbs; 1 and 2 indicate singular or plural; and type of verb is indicated by N, R, O (direct object not possible, required, or optional). (b) Graph of network predictions following presentation of the word **boys**.

But in fact, the prediction (4d) is correctly that the next verb should be in the singular in order to agree with the first noun. In so doing, it has found some mechanism for representing the long-distance dependency between the main clause noun and main clause verb, despite the presence of an intervening noun and verb (with their own agreement relations) in the relative clause.

Note that this sentence also illustrates the sensitivity to an interaction between verb argument structure and relative clause structure. The verb **chases** takes an obligatory direct object. In simple sentences the direct object follows the verb immediately; this is also true

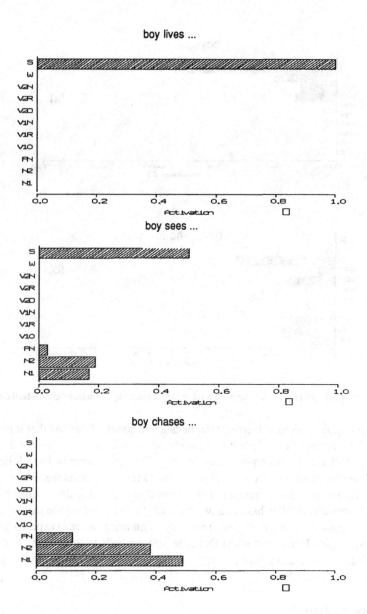

Figure 3. Graph of network predictions following the sequences **boy lives** ...; **boy sees** ...; and **boy chases** ... (the first precludes a direct object, the second optional permits a direct object, and the third requires a direct object).

in many complex sentences (e.g., **boys who chase mary feed cats**). In the sentence displayed, however, the direct object (**boys**) is the head of the relative clause and appears before the verb. This requires that the network learn (a) that there are items which function as nouns, verbs, etc.; (b) which items fall into which classes; (c) that there are subclasses of verbs which have different cooccurrence relations with nouns, corresponding to verb-direct object restrictions; (d) which verbs fall into which classes; and (e) when to expect that the direct object will follow the verb, and when to know that it has already appeared.

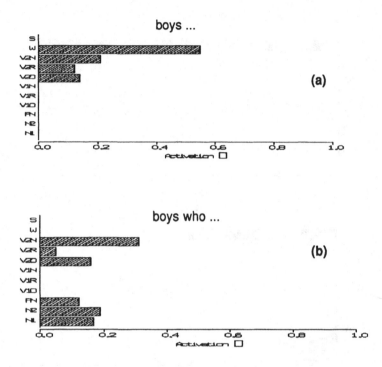

Figure 4. Graph of network predictions after each word in the sentence **boys who mary chases feed dogs.** is input.

The network appears to have learned this, because in panel (d) we see that it expects that **chases** will be followed by a verb (the main clause verb, in this case) rather than a noun.

An even subtler point is demonstrated in (4c). The appearance of **boys** followed by a relative clause containing a different subject (**who Mary. . .**) primes the network to expect that the verb which follows must be of the class that requires a direct object, precisely because a direct object filler has already appeared. In other words, the network correctly responds to the presence of a filler (**boys**) not only by knowing where to expect a gap (following **chases**); it also learns that when this filler corresponds to the object position in the relative clause, a verb is required which has the appropriate argument structure.

5. Network analysis

The natural question to ask at this point is how the network has learned to accomplish the task. Success on this task seems to constitute *prima facie* evidence for the existence of internal representations which possessed abstract structure. That is, it seemed reasonable to believe that in order to handle agreement and argument structure facts in the presence of relative clauses, the network would be required to develop representations which reflected constituent structure, argument structure, grammatical category, grammatical relations, and number. (At the very least, this is the same sort of inference which is made in the case of human language users, based on behavioral data.)

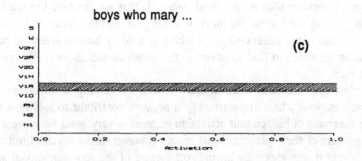

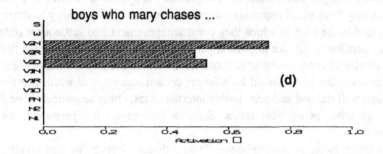

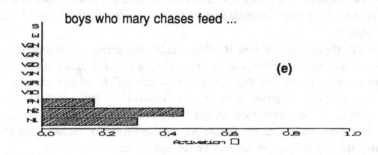

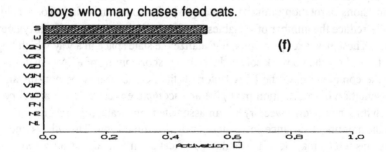

Figure 4. Continued.

One advantage of working with an artificial system is that we can take the additional step of directly inspecting the internal mechanism which generates the behavior. Of course, the mechanism we find is not necessarily that which is used by human listeners; but we may nonetheless be surprised to find solutions to the problems which we might not have guessed on our own.

Hierarchical clustering has been a useful analytic tool for helping to understand how the internal representations which are learned by a network contribute to solving a problem. Clustering diagrams of hidden unit activation patterns is very good for representing the similarity structure of the representational space. However, it has certain limitations. One weakness is that it provides only an indirect picture of the representational space. Another shortcoming is that it tends to deemphasize the dynamics involved in processing. Some states may have significance not simply in terms of their similarity to other states, but with regard to the ways in which they constrain movement into subsequent state space (recall the examples in (1)). An important part of what the network has learned lies in the dynamics involved in processing word sequences. Indeed, one might think of the network dynamics as encoding grammatical knowledge; certain sequences of words move the network through well-defined and permissible internal states. Other sequences move the network through other permissible states. Some sequences are not permitted; these are ungrammatical.

What we might therefore wish to be able to do is directly inspect the internal states (represented by the hidden unit activation vectors) the network is in as it processes words in sequence, in order to see how the states and the trajectories encode the network's grammatical knowledge.

Unfortunately, the high dimensionality of the hidden unit activation vectors (in the simulation here, 70 dimensions) makes it impractical to view the state space directly. Furthermore, there is no guarantee that the dimensions which will be of interest to us—in the sense that they pick out regions of importance in network's solution to the task—will be correlated with any of the dimensions coded by the hidden units. Indeed, this is what it means for the representations to be distributed: the dimensions of variation cut across, to some degree, the dimensions picked out by the hidden units.

However, it is reasonable to assume that such dimensions of variation do exist, we can try to identify them using principal component analysis (PCA). PCA allows us to find another set of dimensions (a rotation of the axes) along which maximum variation occurs.[1] (It may additionally reduce the number of variables by effectively removing the linearly dependent set of axes.) These new axes permit us to visualize the state space in a way which hopefully allows us to see how the network solves the task. (A shortcoming of PCA is that it is linear; however, the combination of the PCA factors at the next level may be non-linear, and so this representation of information may give an incomplete picture of the actual computation.) Each dimension (eigenvector) has an associated eigenvalue, the magnitude of which indicates the amount of variance accounted for by that dimension. This allows one to focus on dimensions which may be of particular significance; it also allows a *post hoc* estimate of the number of hidden units which might actually be required for the task. Figure 5 shows a graph of the eigenvalues of the 70 eigenvectors which were extracted.

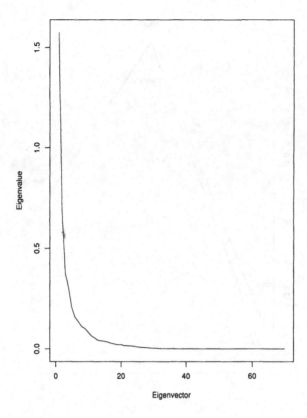

Figure 5. Graph of eigenvalues of the 70 ordered eigenvectors extracted in Simulation 2.

5.1. Agreement

The sentences in (8) were presented to the network, and the hidden unit patterns captured after each word was processed in sequence.

(8a) boys hear boys.
(8b) boy hears boys.
(8c) boy who boys chase chases boy.
(8d) boys who boys chase chase boy.

(These sentences were chosen to minimize differences due to lexical content and to make it possible to focus on differences to grammatical structure. (8a) and (8b) were contained in the training data; (8c) and (8d) were novel and had never been presented to the network during learning.)

By examining the trajectories through state space along various dimensions, it was apparent that the second principal component played an important role in marking number of the main clause subject. Figure 6 shows the trajectories for (8a) and (8b); the trajectories

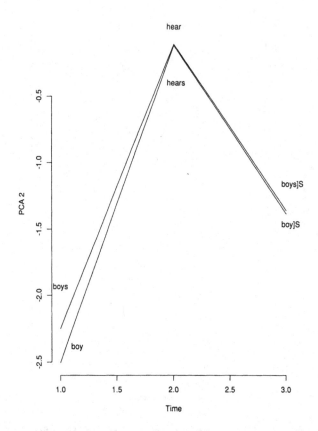

Figure 6. Trajectories through state space for sentences (8a) and (8b). Each point marks the position along the second principle component of hidden units space, after the indicated word has been input. Magnitude of the second principle component is measured along the ordinate; time (i.e., order of word in sentence) is measured along the abscissa. In this and subsequent graphs the sentence-final word is marked with a]S.

are overlaid so that the differences are more readily seen. The paths are similar and diverge only during the first word, indicating the difference in the number of the initial noun. The difference is slight and is eliminated after the main (i.e., second **chase**) verb has been input. This is apparently because, for these two sentences (and for the grammar), number information does not have any relevance for this task once the main verb has been received.

It is not difficult to imagine sentences in which number information may have to be retained over an intervening constituent; sentences (8c) and (8d) are such examples. In both these sentences there is an identical relative clause which follows the initial noun (which differs with regard to number in the two sentences). This material, **who boys chase**, is irrelevant as far as the agreement requirements for the main clause verb. The trajectories through state space for these two sentences have been overlaid and are shown in Figure 7; as can be seen, the differences in the two trajectories are maintained until the main clause verb is reached, at which point the states converge.

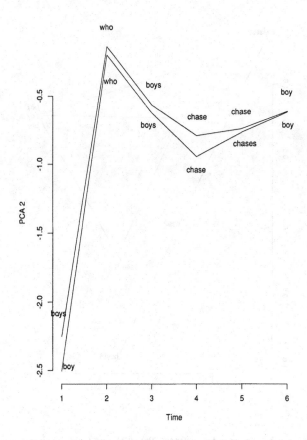

Figure 7. Trajectories through state space during processing of (8c) and (8d).

5.2. Verb argument structure

The representation of verb argument structure was examined by probing with sentences containing instances of the three different classes of verbs. Sample sentences are shown in (9).

(9a) boy walks.
(9b) boy sees boy.
(9c) boy chases boy.

The first of these contains a verb which may not take a direct object; the second takes an option direct object; and the third requires a direct object. The movement through state space as these three sentences are processed are shown in Figure 8.

This figure illustrates how the network encodes severl aspects of grammatical structure. Nouns are distinguished by role; subject nouns for all three sentences appear in the upper right portion of the space, and object nouns appear below them. (Principal component 4, not shown here, encodes the distinction between verbs and nouns., collapsing across case.)

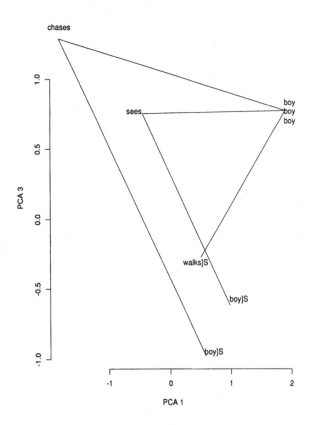

Figure 8. Trajectories through state space for sentences (9a), (9b), and (9c). Principle component 1 is plotted along the abscissa; principal component 3 is plotted along the ordinate.

Verbs are differentiated with regard to their argument structure. **Chases** requires a direct object, **sees** takes an optional direct object, and **walks** precludes an object. The difference is reflected in a systematic displacement in the plane of principal components 1 and 3.

5.3. Relative clauses

The presence of relative clauses introduces a complication into the grammar, in that the representations of number and verb argument structure must be clause-specific. It would be useful for the network to have some way to represent the constituent structure of sentences.

The trained network was given the following sentences.

(10a) boy chases boy.
(10b) boy chases boy who chases boy.
(10c) boy who chases boy chases boy.
(10d) boy chases boy who chases boy who chases boy.

The first sentence is simple; the other three are instances of embedded sentences. Sentence 10a was contained in the training data; sentences 10b, 10c, and 10d were novel and had not been presented to the network during the learning phase.

The trajectories through state space for these four sentences (principal components 1 and 11) are shown in Figure 9. Panel (9a) shows the basic pattern associated with what is in fact the matrix sentences for all four sentences. Comparison of this figure with panels (9b) and (9c) shows that the trajectory for the matrix sentence appears to follow the same for; the matrix subject noun is in the lower left region of state space, the matrix verb appears above it and to the left, and the matrix object noun is near the upper middle region. (Recall that we are looking at only 2 of the 70 dimensions; along other dimensions the noun/verb distinction is preserved categorically.) The relative clause appears involve a replication of this basic pattern, but displaced toward the left and moved slightly downward, relative to the matrix constituents. Moreover, the exact position of the relative clause elements indicates which of the matrix nouns are modified. Thus, the relative clause modifying the subject noun is closer to it, and the relative clause modifying the object noun are closer to it. This trajectory pattern was found for all sentences with the same grammatical form; the pattern is thus systematic.

Figure (9d) shows what happens when there are multiple levels of embedding. Successive embeddings are represented in a manner which is similar to the way that the first embedded clause is distinguished from the main clause; the basic pattern for the clause is replicated in region of state space which is displaced from the matrix material. This displacement provides a systematic way for the network to encode the depth of embedding in the current state. However, the reliability of the encoding is limited by the precision with which states are represented, which in turn depends on factors such as the number of hidden units and the precision of the numerical values. In the current simulation, the representation degraded after about three levels of embedding. The consequences of this degradation on performance (in the prediction task) are different for different types of sentences. Sentences involving center embedding (e.g., 9c and 9d), in which the level of embedding is crucial for maintaining correct agreement, are more adversely affected than sentences involving so-called tail-recursion (e.g., 10d). In these latter sentences the syntactic structures in principle involve recursion, but in practice the level of embedding is not relevant for the task (i.e., does not affect agreement or verb argument structure in any way).

Figure 9d is interesting in another respect. Given the nature of the prediction task, it is actually not necessary for the network to carry forward any information from prior clauses. It would be sufficient for the network to represent each successive relative clause as an iteration of the previous pattern. Yet the two relative clauses are differentiated. Similarly, Servan-Schreiber, Cleeremans, & McClelland (in press) found that when a simple recurrent network was taught to predict inputs that had been generated by a finite state automaton, the network developed internal representations which corresponded to the FSA states; however, it also redundantly made finer-grained distinctions which encoded the path by which the state had been achieved, even though this information was not used for the task. It thus seems to be a property of these networks that while they are able to encode state in a way which minimizes context as far as behavior is concerned, their nonlinear nature allows them to remain sensitive to context at the level of internal representation.

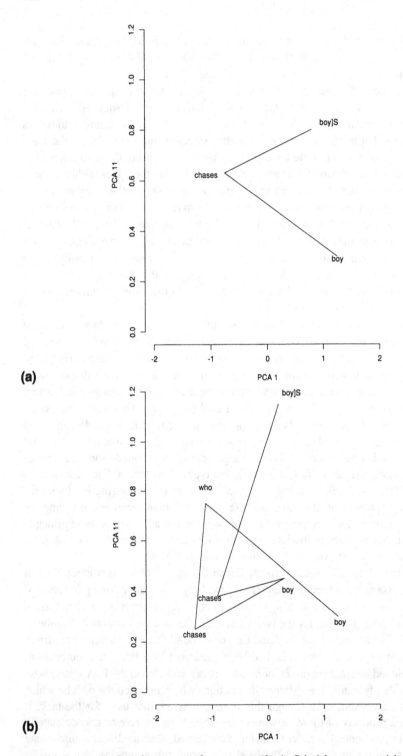

Figure 9. Trajectories through state space for sentences (10a–d). Principle component 1 is displayed along the abscissa; principle component 11 is plotted along the ordinate.

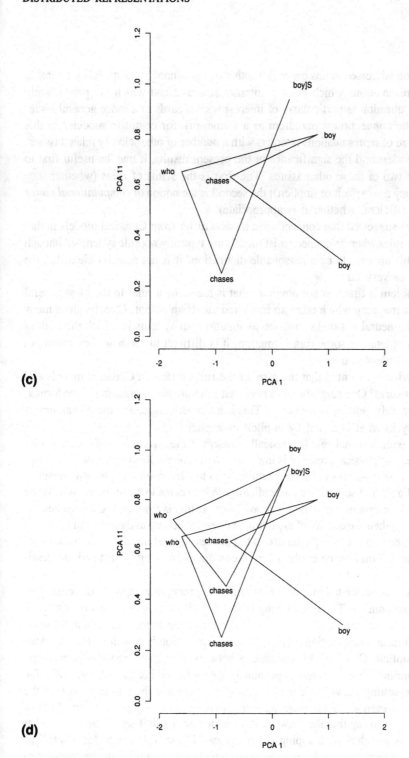

(c)

(d)

Figure 9. Continued.

6. Discussion

The basic question addressed in this paper is whether or not connectionist models are capable of complex representations which possess internal structure and which are productively estensible. This question is particularly of interest with regards to a more general issue: How useful is the connectionist paradigm as a framework for cognitive models? In this context, the nature of representations interacts with a number of other closely related issues. So in order to understand the significance of the present results, it may be useful first to consider briefly two of these other issues. The first is the status of *rules* (whether they exist, whether they are explicit or implicit); the second is the notion of *computational power* (whether it is sufficient, whether it is appropriate).

It is sometimes suggested that connectionist models differ from Classical models in that the latter rely on rules whereas connectionist models are typically not rule systems. Although at first glance this appears to be a reasonable distinction, it is not actually clear that the distinction gets us very far.

The basic problem is that it is not obvious what is meant by a rule. In the most general sense, a rule is a mapping which takes an input and yields an output. Clearly, since many (although not all) neural networks function as input/output systems in which the bulk of the machinery implements some transformation, it is difficult to see how they could not be thought of as rule-systems.

But perhaps what is meant is that the *form* of the rules differs in Classical models and connectionist networks? One suggestion has been that rules are stated *explicitly* in the former, whereas they are only *implicit* in networks. This is a slippery issue, and there is an unfortunate ambiguity in what is meant by implicit or explicit.

One sense of explicit is that rule is physically present in the system *in its form as a rule*; and furthermore, that physical presence is important to the correct functioning of the system. However, Kirsh (1989) points out that our intuitions as to what counts as physical presence are highly unreliable and sometimes contradictory. What seems to really be at stake is the speed with which information can be made available. If this is true, and Kirsh argues the point persuasively, then the quality of explicitness does not belong to data structures alone. One must also take into account the nature of the processing system involved, since information in the same form may be easily accessible in one processing system and inaccessible in another.

Unfortunately, our understanding of the information processing capacity of neural networks is quite preliminary. There is a strong tendency in analyzing such networks to view them through traditional lenses. We suppose that if information is not contained in the same form as more familiar computational systems, that information is somehow buried, inaccessible, and implicit. Consider, for instance, a network which successfully learns some complicated mapping—say, from text to pronunciation (Sejnowski & Rosenberg, 1987). On inspecting the resulting network, it is not immediately obvious how to explain how the mapping works or even to characterize what the mapping is in any precise way. In such cases, it is tempting to say that the network has learned an implicit set of rules. But what we really mean is just that the mapping is "complicated," or "difficult to formulate," or even "unknown." This is rather a description of our own failure to understand the mechanism rather than a description of the mechanism itself. What is needed are new techniques for network analysis, such as the principal component analysis used in the present work, contribution analysis (Sanger, 1989), weight matrix decompositon (McMillan & Smolensky, 1988), or skeletonization (Mozer & Smolensky, 1989).

If successful, these analyses of connectionist networks may provide us with a new vocabulary for understanding information processing. We may learn new ways in whch information can be explicit or implicit, and we may learn new notations for expressing the rules that underlie cognition. The notation of these new connectionist rules may look very different than that used in, for example, production rules. And we may expect that the notation will not lend itself to describing all types of regularity with equal facility.

Thus, the potential important difference between connectionist models and Classical models will not be in whether one or the other systems contains rules, or whether one system encodes information explcitly and the other encodes it implicitly; the difference will lie in the nature of the rules, and in what kinds of information count as explicitly present.

This potential difference brings us to the second issue: computational power. The issue divides into two considerations. Do connectionist models provide *sufficient* computational power (to account for cognitive phenomena); and do they provide the *appropriate* sort of computational power?

The first question can be answered affirmatively with an important qualification. It can be shown that multilayer feedforward networks with as few as one hidden layer, with no squashing at the output and an arbitrary nonlinear activation function at the hidden layer, are capable of arbitrarily accurate approximation of arbitrary mappings. They thus belong to a class of universal approximators (Hornik, Stinchcombe, & White, in press; Stinchcombe & White, 1989). Pollack (1988) has also proven the Turing equivalence of neural networks. In principle, then, such networks are capable of implementing any function that the Classical system can implement.

The important qualification to the above results is that sufficiently many hidden units be provided (or in the case of Pollack's proof, that weights be inifinite precision). What is not currently known is effect of limited resources on computational power. Since human cognition is carried out in a system with relatively fixed and limited resources, this question is of paramount interest. These limitations provide critical constraints on the nature of the functions which can be mapped; it is an important empirical question whether these constraints explain the specific form of human cognition.

It is in this context that the question of the appropriateness of the computational power becomes interesting. Given limited resources, it is relevant to ask whether the kinds of operations and representations which are naturally made available are those which are likely to figure in human cognition. If one has a theory of cognition which requires sorting of randomly ordered information, e.g., word frequency lists in Forster's (1979) model of lexical access, then it becomes extremely important that the computational framework provide efficient support for the sort operation. On the other hand, if one believes that information is stored associatively, then the ability of the system to do a fast sort is irrelevant. Instead, it is important that the model provide for associative storage and retrieval.[2] Of course, things work in both directions. The availability of certain types of operations may encourage one to build models of a type which are impractical in other frameworks. And the need to work with an inappropriate computational mechanism may blind us from seeing things as they really are.

Let us return now to the current work. I would like to discuss first some of the ways in which the work is preliminary and limited. Then I will discuss what I see as the positive contributions of the work. Finally, I would like to relate this work to other connectionist research and to the general question raised at the outset of this discussion: How viable are connectionist models for understanding cognition?

The results are preliminary in a number of ways. First, one can imagine a number of additional tests that could be performed to test the representational capacity of the simple recurrent network. The memory capacity remains unprobed (but see Servan-Schreiber, Cleeremans, & McClelland, in press). Generalization has been tested in a limited way (many of the tests involved novel sentences), but one would like to know whether the network can inferentially extend what it knows about the types of noun phrases encountered in the second simulation (simple nouns and relative clauses) to noun phrases with different structures.

Second, while it is true that the agreement and verb argument structure facts contained in the present grammar are important and challenging, we have barely scratched the surface in terms of the richness of linguistic phenomena which characterize natural languages.

Third, natural languages not only contain far more complexity with regard to their syntactic structure, they also have a semantic aspect. Indeed, Langacker (1987) and others have argued persuasively that it is not fruitful to consider syntax and semantics as autonomous aspects of language. Rather, the form and meaning of language are closely entwined. Although there may be things which can be learned by studying artificial languages such as the present one which are purely syntactic, *natural* language processing is crucially an attempt to retrieve meaning from linguistic form. The present work does not addrress this issue at all, but there are other PDP models which have made progress on this problem (e.g., St. John & McClelland, in press).

What the current work does contribute is some notion of the representational capacity of connectionist models. Various writers (e.g., Fodor & Pylyshyn, 1988) have expressed concern regarding the ability of connectionist representations to encode compositional structure and to provide for open-ended generative capacity. The networks used in the simulations reported here have two important properties which are relevant to these concerns.

First, the networks make possible the development of internal representations that are *distributed* (Hinton, 1988; Hinton, McClelland, Rumelhart, 1986). While not unbounded, distributed representations are less rigidly coupled with resources than localist representations, in which there is a strict mapping between concept and individual nodes. There is also greater flexibility in determining the dimensions of importance for the model.

Second, the networks studied here build in a sensitivity to context. The important result of the current work is to suggest that the sensitivity to context which is characteristic of many connectionist models, and which is built-in to the architecture of the networks used here, does not preclude the ability to capture generalizations which are at a high level of abstraction. Nor is this a paradox. Sensitivity to context is precisely the mechanism which underlies the ability to abstract and generalize. The fact that the networks here exhibited behavior which was highly regular was not because they learned to be context-insensitive. Rather, they learned to respond to contexts which are more abstractly defined. Recall that even when these networks' behavior seems to ignore context (e.g., Figure 9d; and Servan-Schreiber, Cleeremans, & McClelland, in press), the internal representations reveal that contextual information is still retained.

This behavior is in striking contrast to that of traditional symbolic models. Representations in these systems are naturally context-*insensitive*. This insensitivity makes it possible to express generalizations which are fully regular at the highest possible level of representation (e.g., purely syntactic), but they require additional apparatus to account for regularities which reflect the interaction of meaning with form and which are more contextually defined. Connectionist models on the other hand begin the task of abstraction at the other end of the continuum. They emphasize the importance of context and the interaction of form with meaning. As the current work demonstrates, these characteristics lead quite naturally to generalizations at a high level of abstraction where appropriate, but the behavior remains ever-rooted in representations whch are contextually grounded. The simulations reported here do not capitalize on subtle distinctions in context, but there are ample demonstrations of models which do (e.g., Kawamoto, 1988; McClelland & Kawamoto, 1986; Miikkulainen & Dyer, 1989; St. John & McClelland, in press).

Finally, I wish to point out that the current approach suggests a novel way of thinking about how mental representations are constructed from language input.

Conventional wisdom holds that as words are heard, listeners retrieve lexical representations. Although these representations may indicate the contexts in which the words acceptably occur, the representations are themselves context-free. They exist in some canonical form which is constant across all occurrences. These lexical forms are then used to assist in constructing a complex representation into which the forms are inserted. One can imagine that when complete, the result is an elaborate structure in which not only are the words visible, but which also depicts the abstract grammatical structure which binds those words.

In this account, the process of building mental structures is not unlike the process of building any other physical structure, such as bridges or houses. Words (and whatever other representational elements are involved) play the role of building blocks. As is true of bridges and houses, the building blocks are themselves unaffected by the process of construction.

A different image is suggested in the approach taken here. As words are processed there is no separate stage of lexical retrieval. There are no representations of words in isolation. The representations of words (the internal states following input of a word) always reflect the input taken together with the prior state. In this scenario, words are not building blocks as much as they are cues which guide the network through different grammatical states. Words are distinct from each other by virtue of having different causal properties.

A metaphor which captures some of the characteristics of this approach is the combination lock. In this metaphor, the role of words is analogous to the role played by the numbers in the combination. The numbers have causal properties; they advance the lock into different states. The effect of a number is dependent on its context. Entered in the correct sequence, the numbers move the lock into an open state. The open state may be said to be *functionally compositional* (van Gelder, in press) in the sense that it reflects a particular sequence of events. The numbers are "present" insofar as they are responsible for the final state, but not because they are still physically present.

The limitation of the combination lock is of course that there is only one correct combination. The networks studied here are more complex. The causal properties of the words are highly structure-dependent and the networks allow many "open" (i.e., grammatical) states.

This view of language comprehension emphasizes the functional importance of representations and is similar in spirit to the approach described in Bates & MacWhinney, 1982; McClelland, St. John, & Taraban, 1989; and many others who have stressed the functional nature of language. Representations of language are constructed in order to accomplish some behavior (where, obviously, that behavior may range from day-dreaming to verbal duels, and from asking directions to composing poetry). The representations are not propositional, and their information content changes constantly over time in accord with the demands of the current task. Words serve as guideposts which help establish mental states that support this behavior; representations are snapshots of those mental states.

Acknowledgments

I am grateful for many useful discussions on this topic with Jay McClelland, Dave Rumelhart, Elizabeth Bates, Steve Stich, and members of the UCSD PDP/NLP Research Group. I thank McClelland, Mike Jordan, Mary Hare, Ken Baldwin, and two anonymous reviewers for critical comments on earlier versions of this paper. This research was supported by contracts N00014-85-K-0076 from the Office of Naval Research and contract DAAB-07-87-C-H027 from Army Avionics, Ft. Monmouth. Requests for reprints should be sent to the Center for Research in Language, 0126; University of California, San Diego; La Jolla, CA 92093-0126.

Notes

1. In practical terms, this analysis involves passing the training set through the trained network (with weights frozen) and saving the hidden unit patterns that are produced in response to each input. The covariance matrix of the resulting set of hidden unit vectors is calculated, and then the eigenvectors of the covariance matrix are found. The eigenvectors are ordered by the magnitude of their eigenvalues, and are used as the basis for describing the original hidden unit vectors. This new set of dimensions has the effect of giving a somewhat more localized description to the hidden unit patterns, because the new dimensions now correspond to the location of meaningful activity (defined in terms of variance) in the hyperspace. Since the dimensions are ordered in terms of variance accounted for, we may wish to look at selected dimensions, starting with those with largest eigenvalues. See Flury (1988) for a detailed explanation of PCA; or Gonzalez & Wintz (1977) for a detailed description of the algorithm.
2. This example was suggested to me by Don Norman.

References

Baker, C.L. (1979). Syntactic theory and the projection problem. *Linguistic Inquiry, 10*, 533–581.
Bates, E., & MacWhinney, B. (1982). Functionalist approaches to grammar. In E. Wanners, & L. Gleitman (Eds.), *Language acquisition: The state of the art*. New York: Cambridge University Press.
Chafe, W. (1970). *Meaning and the structure of language*. Chicago: University of Chicago Press.
Chalmers, D.J. (1990). Syntactic transformations on distributed representations. Center for Research on Concepts and Cognition, Indiana University.
Chomsky, N. (1957). *Syntactic structures*. The Hague: Mouton.
Dell, G. (1986). A spreading activation theory of retrieval in sentence production. *Psychological Review, 93*, 283–321.

Dolan, C., & Dyer, M.G. (1987). *Symbolic schemata in connectionist memories: Role binding and the evolution of structure* (Technical Report UCLA-AI-87-11). Los Angeles, CA: University of California, Los Angeles, Artificial Intelligence Laboratory.

Dolan, C.P., & Smolensky, P. (1988). *Implementing a connectionist production system using tensor products* (Technical Report UCLA-AI-88-15). Los Angeles, CA: University of California, Los Angeles, Artificial Intelligence Laboratory.

Elman, J.L. (1989). *Representation and structure in connectionist models* (Technical Report CRL-8903). San Diego, CA: University of California, San Diego, Center for Research in Language.

Elman, J.L. (1990). Finding structure in time. *Cognitive Science, 14,* 179–211.

Fauconnier, G. (1985) *Mental spaces.* Cambridge, MA: MIT Press.

Feldman, J.A. & Ballard, D.H. (1982). Connectionist models and their properties. *Cognitive Science, 6,* 205–254.

Fillmore, C.J. (1982). Frame semantics. In *Linguistics in the morning calm.* Seoul: Hansin.

Flury, G. (1988). *Common principal components and related multivariate models.* New York: Wiley.

Fodor, J. (1976). *The language of thought.* Harvester Press, Sussex.

Fodor, J., & Pylyshyn, Z. (1988). Connectionism and cognitive architecture: A critical analysis. In S. Pinker & J. Mehler (Eds.), *Connections and symbols.* Cambridge, MA: MIT Press.

Forster, K.I. (1979). Levels of processing and the structure of the language processor. In W.E. Cooper, & E. Walker (Eds.), *Sentence processing: Psycholinguistic studies presented to Merrill Garrett.* Hillsdale NJ: Lawrence Erlbaum Associates.

Gasser, M., & Lee, C-D. (1990). Networks that learn phonology. Computer Science Department, Indiana University.

Givon, T. (1984). *Syntax: A functional-typological introduction. Volume 1.* Amsterdam: John Benjamins.

Gold, E.M. (1967). Language identification in the limit. *Information and Control, 16,* 447–474.

Gonzalez, R.C., & Wintz, P. (1977). *Digital image processing.* Reading, MA: Addison-Wesley.

Grosjean, F. (1980). Spoken word recognition processes and the gating paradigm. *Perception & Psychophysics, 28,* 267–283.

Hanson, S.J., & Burr, D.J. (1987). Knowledge representation in connectionist networks. Bell Communications Research, Morristown, New Jersey.

Hare, M. (1990). *The role of similarity in Hungarian vowel harmony: A connectionist account* (CRL Technical Report 9004). San Diego, CA: University of California, Center for Research in Language.

Hare, M., Corina, D., & Cottrell, G. (1988). *Connectionist perspective on prosodic structure* (CRL Newsletter, Vol. 3, No. 2). San Diego, CA: University of California, Center for Research in Language.

Hinton, G.E. (1988). *Representing part-whole hierarchies in connectionist networks* (Technical Report CRG-TR-88-2). University of Toronto, Connectionist Research Group.

Hinton, G.E., McClelland, J.L., & Rumelhart, D.E. (1986). Distributed representations. in D.E. Rumelhart, & J.L. McClelland (Eds.), *Parallel distributed processing: Explorations in the microstructure of cognition (Vol. 1).* Cambridge, MA: MIT Press.

Hopper, P.J., & Thompson, S.A. (1980). Transitivity in grammar and discourse. *Language, 56,* 251–299.

Hornik, K., Stinchcombe, M., & White, H. (in press). Multi-layer feedforward networks are universal approximators. *Neural Networks.*

Jordan, M.I. (1986). *Serial order: A parallel distributed processing approach* (Technical Report 8604). San Diego, CA: University of California, San Diego, Institute for Cognitive Science.

Kawamoto, A.H. (1988). Distributed representations of ambiguous words and their resolution in a connectionist network. In S.L. Small, G.W. Cottrell, & M.K. Tanenhaus (Eds.), *Lexical ambiguity resolution: Perspectives from psycholinguistics, neuropsychology, and artificial intelligence.* San Mateo, CA: Morgan Kaufmann Publishers.

Kirsh, D. (in press). When is information represented explicitly? In J. Hanson (Ed.), *Information, thought, and content.* Vancouver: University of British Columbia.

Kuno, S. (1987). *Functional syntax: Anaphora, discourse and empathy.* Chicago: The University of Chicago Press.

Kutas, M. (1988). Event-related brain potentials (ERPs) elicited during rapid serial presentation of congruous and incongruous sentences. In R. Rohrbaugh, J. Rohrbaugh, & P. Parasuramen (Eds.), *Current trends in brain potential research (EEG Supplement 40).* Amsterdam: Elsevier.

Kutas, M., & Hillyard, S.A. (1980). Reading senseless sentences: Brain potentials reflect semantic inconguity. *Science, 207,* 203–205.

Lakoff, G. (1987). *Women, fire, and dangerous things: What categories reveal about the mind.* Chicago: University of Chicago Press.

Langacker, R.W. (1987). *Foundations of cognitive grammar: Theoretical perspectives. Volume 1*. Stanford: Stanford University Press.

Langacker, R.W. (1988). A usage-based model. *Current Issues in Linguistic Theory, 50*, 127-161.

MacWhinney, B., Leinbach, J., Taraban, R., & McDonald, J. (1989). Language learning: Cues or rules? *Journal of Memory and Language, 28*, 255-277.

Marslen-Wilson, W., & Tyler, L.K. (1980). The temporal structure of spoken language understanding. *Cognition, 8*, 1-71.

McClelland, J.L. (1987). The case for interactionism in language processing. In M. Coltheart (Ed.), *Attention and performance XII: The psychology of reading*. London: Erlbaum.

McClelland, J.L., St. John, M., & Taraban, R. (1989). *Sentence comprehension: A parallel distributed processing approach*. Manuscript, Department of Psychology, Carnegie Mellon University.

McMillan, C., & Smolensky, P.)1988). *Analyzing a connectionist model as a system of soft rules* (Technical Report CU-CS-303-88). University of Colorado, Boulder, Department of Computer Science.

Miikkulainen, R., & Dyer, M. (1989a). Encoding input/output representations in connectionist cognitive systems. In D.S. Touretzky, G.E. Hinton, & T.J. Sejnowski (Eds.), *Proceedings of the 1988 Connectionist Models Summer School*. Los Altos, CA: Morgan Kaufmann Publishers.

Miikkulainen, R., & Dyer, M. (1989b). A modular neural network architecture for sequential paraphrasing of script-based stories. In *Proceedings of the International Joint Conference on Neural Networks*, IEEE.

Mozer, M. (1988). *A focused back-propagation algorithm for temporal pattern recognition*. (Technical Report CRG-TR-88-3). University of Toronto, Departments of Psychology and Computer Science.

Mozer, M.C., & Smolensky, P. (1989). *Skeletonization: A technique for trimming the fat from a network via relevance assessment* (Technical Report CU-CS-421-89). University of Colorado, Boulder, Department of Computer Science.

Oden, G. (1978). Semantic constraints and judged preference for interpretations of ambiguous sentences. *Memory and Cognition, 6*, 26-37.

Pinker, S. (1989). *Learnability and cognition: The acquisition of argument structure*. Cambridge, MA: MIT Press.

Pollack, J.B. (1988). Recursive auto-associative memory: Decising compositional distributed representations. *Proceedings of the Tenth Annual Conference of the Cognitive Science Society*. Hillsdale, NJ: Lawrence Erlbaum.

Pollack, J.B. (in press). Recursive distributed representations. *Artificial Intelligence*.

Ramsey, W. (1989). *The philosophical implications of connectionism*. Ph.D. thesis, University of California, San Diego.

Reich, P.A., & Dell, G.S. (1977). Finiteness and embedding. In E.L. Blansitt, Jr., & P. Maher (Eds.), *The third LACUS forum*. Columbia, SC: Hornbeam Press.

Rumelhart, D.E., Hinton, G.E., & Williams, R.J. (1986). Learning internal representations by error propagation. In D.E. Rumelhart, & J.L. McClelland (Eds.), *Parallel distributed processing: Explorations in the microstructure of cognition (Vol. 1)*. Cambridge, MA: MIT Press.

Rumelhart, D.E., & McClelland, J.L. (1986a). PDP Models and general issues in cognitive science. In D.E. Rumelhart, & J.L. McClelland (Eds.), *Parallel distributed processing: Explorations in the microstructure of cognition (Vol. 1)*. Cambridge, MA: MIT Press.

Rumelhart, D.E., & McClelland, J.L. (1986b). On learning the past tenses of English verbs. In D.E. Rumelhart, & J.L. McClelland (Eds.), *Parallel distributed processing: Explorations in the microstructure of cognition (Vol. 1)*. Cambridge, MA: MIT Press.

Salasoo, A., & Pisoni, D.B. (1985). Interaction of knowledge sources in spoken word identification. *Journal of Memory and Language, 24*, 210-231.

Sanger, D. (1989). *Contribution analysis: A technique for assigning responsibilities to hidden units in connectionist networks* (Technical Report CU-CS-435-89). University of Colorado, Boulder, Department of Computer Science.

Schlesinger, I.M. (1971). On linguistic competence. IN Y. Bar-Hillel (Ed.), *Pragmatics of natural languages*. Dordrecht, Holland: Reidel.

Sejnowski, T.J., & Rosenberg, C.R. (1987). Parallel networks that learn to pronounce English text. *Complex Systems, 1*, 145-168.

Servan-Schreiber, D., Cleeremans, A., & McClelland, J.L. (1991). Graded state machines: The representation of temporal contingencies in simple recurrent networks. *Machine Learning, 7*, 161-193.

Shastri, L., & Ajjanagadde, V. (1989). *A connectionist system for rule based reasoning with multi-place predicates and variables* (Technical Report MS-CIS-8905). University of Pennsylvania, Computer and Information Science Department.

Smolensky, P. (1987a). *On variable binding and the representation of symbolic structures in connectionist systems* (Technical Report CU-CS-355-87). University of Colorado, Boulder, Department of Computer Science.

Smolensky, P. (1987b). *On the proper treatment of connectionism* (Technical Report CU-CS-377-87). University of Colorado, Boulder, Department of Computer Science.

Smolensky, P. (1987c). *Putting together connectionism—again* (Technical Report CU-CS-378-87). University of Colorado, Boulder, Department of Computer Science.

Smolensky, P. (1988). On the proper treatment of connectionism. *The Behavioral and Brain Sciences, 11.*

Smolensky, P. (in press). Tensor product variable binding and the representation of symbolic structures in connectionist systems. *Artificial Intelligence.*

St. John, M., & McClelland, J.L. (in press). *Learning and applying contextual constraints in sentence comprehension* (Technical Report). Pittsburgh, PA: Carnegie Mellon University, Department of Psychology.

Stemberger, J.P. (1985). *The lexicon in a model of language production.* New York: Garland Publishing.

Stinchcombe, M., & White, H. (1989). Universal approximation using feedforward networks with non-sigmoid hidden layer activation functions. *Proceedings of the International Joint Conference on Neural Networks,* Washington, D.C.

Stolz, W. (1967). A study of the ability to decode grammatically novel sentences. *Journal of Verbal Learning and Verbal Behavior, 6,* 867–873.

Tanenhaus, M.K., Garnseyh, S.M., & Boland, J. (in press). Combinatory lexical information and language comprehension. In G. Altmann (Ed.), *Cognitive models of speech processing: Psycholinguistic and computational perspectives.* Cambridge, MA: MIT Press.

Touretzky, D.S. (1986). BoltzCONS: Reconciling connectionism with the recursive nature of stacks and trees. *Proceedings of the Eight Annual Conference of the Cognitive Science Society.* Hillsdale, NJ: Lawrence Erlbaum.

Touretzky, D.S. (1989). *Rules and maps in connectionist symbol processing* (Technical Report CMU-CS-89-158). Pittsburgh, PA: Carnegie Mellon University, Department of Computer Science.

Touretzky, D.S. (1989). Towards a connectionist phonology: The "many maps" approach to sequence manipulation. *Proceedings of the 11th Annual Conference of the Cognitive Science Society,* 188–195.

Touretzky, D.S., & Hinton, G.E. (1985). Symbols among the neurons: Details of a connectionist inference architecture. *Proceedings of the Ninth International Joint Conference on Artificial Intelligence, Los Angeles.*

Touretzky, D.S., & Wheeler, D.W. (1989). *A connectionist implementation of cognitive phonology* (Technical Report CMU-CS-89-144). Pittsburgh, PA: Carnegie Mellon University, School of Computer Science.

Van Gelder, T.J. (in press). Compositionality: Variations on a classical theme. *Cognitive Science.*

Machine Learning, 7, 227–252 (1991)
© 1991 Kluwer Academic Publishers, Boston. Manufactured in The Netherlands.

The Induction of Dynamical Recognizers

JORDAN B. POLLACK   pollack@cis.ohio-state.edu
Laboratory for AI Research & Computer & Information Science Department, The Ohio State University,
2036 Neil Avenue, Columbus, OH 43210

Abstract. A higher order recurrent neural network architecture learns to recognize and generate languages after being "trained" on categorized exemplars. Studying these networks from the perspective of dynamical systems yields two interesting discoveries: First, a longitudinal examination of the learning process illustrates a new form of mechanical inference: Induction by phase transition. A small weight adjustment causes a "bifurcation" in the limit behavior of the network. This phase transition corresponds to the onset of the network's capacity for generalizing to arbitrary-length strings. Second, a study of the automata resulting from the acquisition of previously published training sets indicates that while the architecture is *not* guaranteed to find a minimal finite automaton consistent with the given exemplars, which is an NP-Hard problem, the architecture does appear capable of generating non-regular languages by exploiting fractal and chaotic dynamics. I end the paper with a hypothesis relating linguistic generative capacity to the behavioral regimes of non-linear dynamical systems.

Keywords. Connectionism, language, induction, dynamics, fractals

1. Introduction

Consider the two categories of binary strings in Table 1. After a brief study, a human or machine learner might decide to characterize the "accept" strings as those containing an odd number of 1's and the "reject" strings as those containing an even number of 1's.

The language acquisition problem has been around for a long time. In its narrowest formulation, it is a version of the inductive inference or "theory from data" problem for syntax: Discover a compact mathematical description of string acceptability (which generalizes) from a finite presentation of examples. In its broadest formulation it involves accounting for the psychological and linguistic facts of native language acquisition by human children, or even the acquistion of language itself by *Homo Sapiens* through natural selection (Lieberman, 1984; Pinker & Bloom, 1990).

Table 1. What is the rule which defines the language?

Accept	Reject
1	0
0 1	0 0
1 0	1 1
1 0 1 1 0	1 0 1
0 0 1	0 1 0 1
1 1 1	1 0 0 0 1
0 1 0 1 1	

The problem has become specialized across many scientific disciplines, and there is a voluminous literature. Mathematical and computational theorists are concerned with the basic questions and definitions of language learning (Gold, 1967), with understanding the complexity of the problem (Angluin, 1978; Gold, 1978), or with good algorithms (Berwick, 1985; Rivest & Schapire, 1987). An excellent survey of this approach to the problem has been written by (Angluin & Smith, 1983). Linguists are concerned with grammatical frameworks which can adequately explain the basic fact that children acquire their language (Chomsky, 1965; Wexler & Culicover, 1980), while psychologists and psycholinguists are concerned, in detail, with how an acquisition mechanism substantiates and predicts empirically testable phenomena of child language acquisition. (MacWhinney, 1987; Pinker, 1984).

My goals are much more limited than either the best algorithm or the most precise psychological model; in fact I scrupulously avoid any strong claims of algorithmic efficiency, or of neural or psychological plausibility for this initial work. I take as a central research question for connectionism:

> *How could a neural computational system, with its slowly-changing structure, numeric calculations, and iterative processes,* ever come to possess *linguistic generative capacity, which seems to require dynamic representations, symbolic computation, and recursive processes?*

Although a rigorous theory may take some time to develop, the work I report in this paper does address this question. I expose a recurrent higher order back-propagation network to both positive and negative examples of boolean strings, and find that although the network does *not* converge on the minimal-description finite state automaton for the data (which is NP-Hard), it does induction in a novel and interesting fashion, and searches through a hypothesis space which, theoretically, is not constrained to machines of finite state.

These results are of import to many related neural models currently under development, e.g., (Elman, 1990; Giles et al., 1990; Servan-Schreiber et al., 1989), and ultimately relates to the question of how linguistic capacity can arise in nature.

Of necessity, I will make use of the terminology of non-linear dynamical systems for the remainder of this article. This terminology is not (yet) a common language to most computer and cognitive scientists and thus warrants an introduction. The view of neural networks as non-linear dynamical systems is commonly held by the physicists who have helped to define the modern field of neural networks (Hopfield, 1982; Smolensky, 1986), although complex dynamics have generally been suppressed in favor of more tractable convergence (limit point) dynamics. But chaotic behavior has shown up repeatedly in studies of neural networks (Derrida & Meir, 1988; Huberman & Hogg, 1987; Kolen & Pollack, 1990; Kurten, 1987; van der Maas et al., 1990), and a few scientists have begun to explore how this dynamical complexity could be exploited for useful purposes, e.g., (Hendin et al., 1991; Pollack, 1989; Skarda & Freeman, 1987).

In short, a discrete dynamical system is just an iterative computation. Starting in some "initial condition" or state, the next state is computed as a mathematical function of the current state, sometimes involving parameters and/or input or noise from an environment. Rather than studying the *function* of the computations, much of the work in this field has been concerned with explaining universal temporal *behaviors*. Indeed, iterative systems have some interesting properties: Their behavior in the limit reaches either a steady state (limit point), an oscillation (limit cycle), or an aperiodic instability (chaos). In terms of computer programs, these three "regimes" correspond, respectively, to those programs which halt, those which have simple repetitive loops, and those which have more "creative" infinite loops, such as broken self-modifying codes, an area of mechanical behavior which has not been extensively studied. When the state spaces of dynamical systems are plotted, these three regimes have characteristic figures called "attractors": Limit points show up as "point attractors," limit cycles as "periodic attractors," and chaos as "strange attractors," which usually have a "fractal" nature. Small changes in controlling parameters can lead through "phase transitions" to these qualitatively different behavioral regimes; a "bifurcation" is a change in the periodicity of the limit behavior of a system, and the route from steady-state to periodic to aperiodic behavior follows a universal pattern. Finally, one of the characteristics of chaotic systems is that they can be very sensitive to initial conditions, and a slight change in the initial condition can lead to radically different outcomes. Further details can be found in articles and books on the field, e.g., (Crutchfield et al., 1986; Devaney, 1987; Gleick, 1987; Grebogi et al., 1987).

2. Automata recurrent networks, and dynamical recognizers

I should make it clear from the outset that the problem of inducing *some* recognizer for a finite set of examples is "easy," as there are an infinite number of regular languages which account for a finite sample, and an infinite number of automata for each language. The difficult problem has always been finding the "minimal description," and no solution is asymptotically much better than "learning by enumeration"—brute-force searching of all automata in order of ascending complexity. Another difficult issue is the determination of grammatical class. Because a finite set of examples does not give any clue as to the complexity class of the source language, one apparently must find the most parsimonious regular grammar, context-free grammar, context-sensitive grammar, etc., and compare them. Quite a formidable challenge for a problem-solver!

Thus, almost all language acquisition work has been done with an inductive bias of presupposing some grammatical framework as the hypothesis space. Most have attacked the problem of inducing finite-state recognizers for regular languages, e.g., (Feldman, 1972; Tomita, 1982).

A Finite State Recognizer is a quadruple $\{Q, \Sigma, \delta, F\}$, where Q is a set of states (q_0 denotes the initial state), Σ is a finite input alphabet, δ is a transition function from $Q \times \Sigma \Rightarrow Q$ and F is a set of final (accepting) states, a subset of Q. A string is accepted by such a device, if, starting from q_0, the sequence of transitions dictated by the tokens in the string ends up in one of the final states.

Table 2. δ function for parity machine.

State	Input	
	0	1
q_0	q_0	q_1
q_1	q_1	q_0

δ is usually specified as a table which lists a new state for each state and input. As an example, a machine which accepts boolean strings of odd parity can be specified as $Q = \{q_0, q_1\}$, $\Sigma = \{0, 1\}$, $F = \{q_1\}$, and δ as shown in table 2.

Although such machines are usually described with fully explicit tables or graphs, transition functions can also be specified as a mathematical function of codes for the current state and the input. For example, variable-length parity can be specifed as the exclusive-or of the current state and the input, each coded as a single bit. The primary result in the field of neural networks is that under simplified assumptions, networks have the capacity to perform arbitrary logical functions, and thus to act as finite-state controllers (McCulloch & Pitts, 1943; Minsky, 1972). In various configurations, modern multilayer feed-forward networks are also able to perform arbitrary boolean functions (Hornik et al., 1990; Lapedes & Farber, 1988; Lippman, 1987). Thus when used recurrently, these networks have the capacity to be any finite state recognizer as well. The states and tokens are assigned binary codes (say with one bit indicating which states are in F), and the code for the next state is simply computed by a set of boolean functions of the codes for current state and current input.

But the mathematical models for neural nets are "richer" than boolean functions, and more like polynomials. What does this mean for automata? In order not to confuse theory and implementation, I will first defne a general mathematical object for language recognition as a forced discrete-time continuous-space dynamical system plus a precise initial condition and a decision function. The recurrent neural network architecture presented in the next section is a constrained implementation of this object.

By analogy to a finite-state recognizer, a *Dynamical Recognizer* is a quadruple $\{Z, \Sigma, \Omega, G\}$, where $Z \subset R^k$ is a "space" of states and $z_k(0)$ is the initial condition. Σ is a finite input alphabet. Ω is the "dynamic," a parameterized set (one for each token) of transformations on the space $\omega_{\sigma_i}:Z \rightarrow Z$, and $G(Z) \rightarrow \{0, 1\}$ is the "decision" function.

Each finite length string of tokens in Σ^*, $\sigma_1, \sigma_2, \ldots \sigma_n$, has a final state associated with it, computed by applying a precise sequence of transformations to the initial state: $z_k(n) = \omega_{\sigma_i}(\ldots(\omega_{\sigma_2}(\omega_{\sigma_1}(z_k(0)))))$. The language accepted and generated[1] by a dynamical recognizer is the set of strings in Σ^* whose final states pass the decision test.

In the "Mealy Machine" formulation (Mealy, 1955), which I use in the model below, the decision function applies to the penultimate state and the final token: $G(z_k(n - 1), \sigma_n) \rightarrow \{0, 1\}$. Just as in the case for finite automata, labeling the arcs rather than the nodes can often result in smaller machines.

There are many variants possible, but both Ω and G must be constrained to avoid the vacuous case where some ω or G is as powerful as a Turing Machine. For purposes of this paper, I will assume that G is as weak as a conventional neural network decision function, e.g., a hyperplane or a convex region, and that each ω is as weak as a linear or quasilinear transformation. G could also be a graded function instead of a forced decision, which would lead to a "more-or-less" notion of string acceptability, or it could be a function which returned a more complex categorization or even a *representation*, in which case I would be discussing dynamical *parsers*. Finally, one could generalize from discrete symbols to continuous symbols (MacLennan, 1989; Touretzky & Geva, 1987), or from discrete-time to continuous-time systems (Pearlmutter, 1989; Pineda, 1987).

There are some difficult questions which can be asked immediately about dynamical recognizers. What kind of languages can they recognize and generate? How does this mathematical description compare to various formal grammars on the grounds of parsimony, efficiency of parsing, neural and psychological plasubility, and learnability? I do not yet have the definitive answers to these questions, as this paper is the first study, but will touch on some of these issues later.

One thing is clear from the outset, that even a linear dynamical recognizer model can function as an arbitrary finite state automaton. The states of the automaton are "embedded" in a finite dimensional space such that a linear transformation can account for the state transitions associated with each token. Consider the case where each of k states is a k-dimensional binary unit vector (a 1-in-k code). Each ω_{σ_i} is simply a permutation matrix which "lists" the state transitions for each token, and the decision function is just a logical mask which selects those states in F. It is perhaps an interesting theoretical question to determine the minimum dimensionality of such a linear "embedding" for an arbitrary regular language.

With the introduction of non-linearities, more complex grammars can also be accounted for. Consider a one-dimensional system where Z is the unit line, $z_0 = 1$, G tests if $z(n) > .75$, and $\Sigma = \{L, R\}$. If the transformation ω_L is "multiply z by 0.5" and ω_R is "multiply z by 2 modulo 2" (which only applies when z(i) is 0 or 1), then the recognizer accepts the balanced parentheses language. In other words, it is just as mathematically possible to embed an "infinite state machine" in a dynamical recognizer as it is to embed a finite state machine. I will return to these issues in the conclusion.

To begin to address the question of learnability, I now present and elaborate upon my earlier work on Cascaded Networks (Pollack, 1987a), which were used in a recurrent fashion to learn parity and depth-limited parenthesis balancing, and to map between word sequences and propositional representations (Pollack, 1990).

3. The model

A Cascaded Network is a well-behaved higher-order (Sigma-Pi) connectionist architecture to which the back-propagation technique of weight adjustment (Rumelhart et al., 1986) can be applied. Basically, it consists of two subnetworks in a master-slave relationship: The *function* (slave) network is a standard feed-forward network, with or without hidden layers. However, the weights on the function network are dynamically computed by the linear *context* (master) network. A context network has as many outputs as there are weights in the function network. Thus the input to the context network is used to "multiplex" the function computed, a divide and conquer heuristic which can make learning easier.

When the outputs of the function network are used as recurrent inputs to the context network, a system can be built which learns to associate specific outputs for variable length input sequences. A block diagram of a Sequential Cascaded Network is shown in Figure 1. Because of the multiplicative connections, each input is, in effect, processed by a different function. Given an initial context, $z_k(0)$ (all .5's by default), and a sequence of inputs,

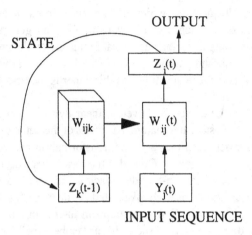

Figure 1. A Sequential Cascaded Network. The outputs of the master net (left) are the weights in the slave net (right), and the outputs of the slave net are recurrent inputs to the master net.

$y_j(t)$, $t = 1 \ldots n$, the network computes a sequence of output/state vectors, $z_i(t)$, $t = 1 \ldots n$ by dynamically changing the set of weights, $w_{ij}(t)$. Without hidden units, the forward-pass computation is:

$$w_{ij}(t) = \sum_k w_{ijk} z_k(t - 1)$$

$$z_i(t) = g(\sum_j w_{ij}(t) y_j(t))$$

which reduces to:

$$z_i(t) = g(\sum_j \sum_k w_{ijk} z_k(t - 1) y_j(t)) \tag{1}$$

where $g(v) = 1/1 + e^{-v}$ is the usual sigmoid function used in back-propagation systems.

In previous work, I assumed that a teacher could supply a consistent and generalizable final-output for each member of a set of strings, which turned out to be a significant over-constraint. In learning a two-state machine like parity, this did not matter, as the 1-bit state fully determines the output. However, for the case of a higher-dimensional systems, we may know what the final output of a system should be, but we *don't care* what its final state is.

Jordan (1986) showed how recurrent back-propagation networks could be trained with "don't care" conditions. If there is no specific target for an output unit during a particular training example, simply consider its error gradient to be 0. This will work, *as long as that same unit receives feedback from other examples*. When the don't-cares line up, the weights to those units will never change. One possible fix, so-called "Back Propagation through time" (Rumelhart et al., 1986), involves a complete unrolling of a recurrent loop and has had only modest success (Mozer, 1988), probably because of conflicts arising from equivalence constraints between interdependent layers. My fix involves a single *backspace*, unrolling the loop only once. For a particular string, this leads to the calculation of only one error term for each weight (and thus no conflict) as follows. After propagating the errors determined on only a subset of the weights from the "acceptance" unit:

$$\frac{\partial E}{\partial w_{aj}(n)} = (z_a(n) - d_a) \, z_a(n) \, (1 - z_a(n)) \, y_j(n)$$

$$\frac{\partial E}{\partial w_{ajk}} = \frac{\partial E}{\partial w_{aj}(n)} \, z_k(n - 1)$$

The error on the remainder of the weights ($\partial E/\partial w_{ijk}$, $i \neq a$) is calculated using values from the penultimate time step:

$$\frac{\partial E}{\partial z_k(n - 1)} = \sum_a \sum_j \frac{\partial E}{\partial w_{ajk}} \frac{\partial E}{\partial w_{aj}(n)}$$

$$\frac{\partial E}{\partial w_{ij}(n - 1)} = \frac{\partial E}{\partial z_i(n - 1)} \, y_j(n - 1)$$

$$\frac{\partial E}{\partial w_{ijk}} = \frac{\partial E}{\partial w_{ij}(n - 1)} \, z_k(n - 2)$$

The schematic for this mode of back propagation is shown in figure 2, where the gradient calculations for the weights are highlighted. The method applies with small variations whether or not there are hidden units in the function or context network, and whether or not the system is trained with a single "accept" bit for desired output, or a larger pattern (representing a tree structure, for example (Pollack, 1990)). The important point is that the gradients connected to a subset of the outputs are calculated directly, but the gradients connected to don't-care recurrent states are calculated one step back in time. The forward and backward calculations are performed over a corpus of variable-length input patterns, and then all the weights are updated. As the overall squared sum of errors approaches 0, the network improves its calculation of final outputs for the set of strings in the training set. At some threshold, for example, when the network responds with above .8 for accept strings, and below .2 for reject strings, training is halted. The network now classifies the training set and can be tested on its generalization to a transfer set.

Unfortunately, for language work, the generalization must be infinite.

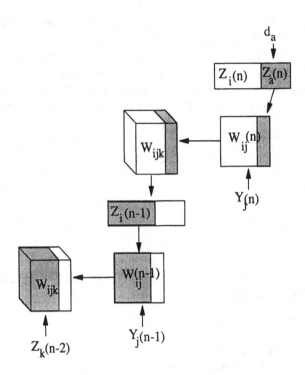

Figure 2. The Backspace Trick. Only partial information is available for computing error gradients on the weights, so the penultimate configuration is used to calculate gradients for the remaining weights.

4. Induction as phase transition

In my original (1987) studies of learning the simple regular language of odd parity, I expected the network to merely implement "exclusive or" with a feedback link. It turns out that this is not quite enough. Because termination of back-propagation is usually defined as a 20% error (e.g., logical "1" is above 0.8), recurrent use of this logic tends to a limit point. In other words, separation of the finite exemplars is no guarantee that the network can recognize sequential parity in the limit. Nevertheless, this is indeed possible as illustrated by the figures below.

A small cascaded network composed of a 1-input 3-output function net (with bias connections, 6 weights for the context net to compute) and a 2-input 6-output context net (with bias connections, 18 weights) was trained an odd parity of a small set of strings up to length 5 (table 1). Of the 3 outputs, two were fed back recurrently as state, and the third was used as the accept unit. At each epoch, the weights in the network were saved in a file for subsequent study. After being trained for about 200 epochs, the network tested successfully on much longer strings. But it is important to show that the network is recognizing parity "in the limit."

In order to observe the limit behavior of a recognizer at various stages of adaptation, we can observe its response to either Σ^* or to a very long "characteristic string" (which

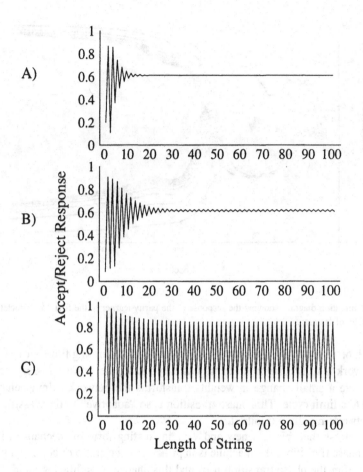

Figure 3. Three stages in the adaptation of a network learning parity. (a) the test cases are separated, but there is a limit point for 1* at about 0.6. (b) after another epoch, the even and odd sequences are slightly separated. (c) after a little more training, the oscillating cycle is pronounced.

has the best chance of breaking it). For parity, a good characteristic string is the sequence of 1's, which should cause the most state changes. Figure 3 shows three stages in the adaptation of a network for parity, by testing the response of three intermediate configurations to the first 100 strings of 1.* In the first figure, despite success at separating the small training set, a single attractor exists in the limit, so that long strings are indistinguishable. After another epoch of training, the even and odd strings are slightly separated, and after still further training, the separation is significant enough to drive a threshold through.

This "phase transition" is shown more completely in figure 4. The vertical axis represents, again, the network's accept/reject response to characteristic strings, but the horizontal axis shows the evolution of this response across all 200 epochs. Each vertical column contains 25 (overlapping) dots marking the network's response to the first 25 characteristic strings. Thus, each "horizontal" line in the graph plots the evolution of the network's response to one of the 25 strings. Initially, all strings longer than length 1 are not distinguished.

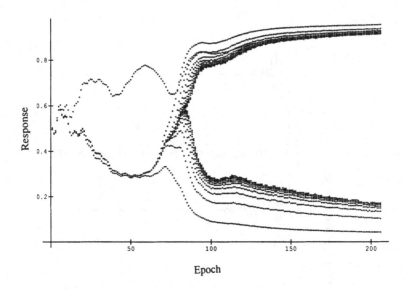

Figure 4. A bifurcation diagram showing the response of the parity-learner to the first 25 characteristic strings over 200 epochs of training.

From epoch 60 to epoch 80, the network is improving at separating finite strings. At epoch 84, the network is still failing in the limit, but at epoch 85, the network undergoes a "bifurcation," where a small change in weights transforms the network's limit behavior from limit point to a limit cycle. This phase transition is so "adaptive" to the classification task that the network rapidly exploits it.

I want to stress that this is a new and very interesting form of mechanical induction. Before the phase transition, the machine is in principle *not* capable of performing the serial parity task; after the phase transition it *is*, and this change in abilities is rapidly exploited by adaptive search. This kind of learning dynamic may be related to biological evolution through natural selection as well as to insight problem-solving (the "aha" phenomenon). The induction is not "one shot" or instantaneous, but more like a "punctuated equilibria" in evolution, where a "pre-adaptive" capacity enables a population some advantage which then drives very rapid change. Metcalfe & Wiebe (1987) report psychological experiments on insight problems in which human subjects measurably undergo a similar cognitive phase transition, reporting no progress on the problems until the solution appears.

5. Benchmarking results

Connectionist and other machine learning algorithms are, unfortunately, very sensitive to the statistical properties of the set of exemplars which make up the learning environment or data-set. When researchers develop their own learning environments, there is a difficult methodological issue bearing on the status of repetitive data-set refinement, especially when experimental results bear on psychologically measured statistics, or the evolution of the data-set is considered too irrelevant to publish. This has correctly led some researchers to include the learning environment as a variable to manipulate (Plunkett & Marchman, 1989). Besides this complicated path, the other methodologically clean choices are to use "real world" noisy data, to choose data once and never refine it, or to use someone else's published training data. For this paper, I chose to use someone else's.

Tomita (1982) performed elegant experiments in inducing finite automata from positive and negative exemplars. He used a genetically inspired two-step hill-climbing procedure, which manipulated 9-state automata by randomly adding, deleting or moving transitions, or inverting the acceptability of a state. Starting with a random machine, the current machine was compared to a mutated machine, and changed only when an improvement was made in the result of a heuristic evaluation function. The first hill-climber used an evaluation function which maximized the difference between the number of positive examples accepted and the number of negative examples accepted. The second hill-climber used an evaluation function which maintained correctness of the examples while minimizing the automaton's description (number of states, then number of transitions) Tomita did not randomly choose his test cases, but instead, chose them consistently with seven regular languages he had in mind (see table 4). The difficulty of these problems lies not in the languages Tomita had in mind, but in the arbitrary and impoverished data sets he used.

Table 3. Training data for seven languages from Tomita (1982).

Set 1 Accept	Set 1 Reject
1	0
1 1	1 0
1 1 1	0 1
1 1 1 1	0 0
1 1 1 1 1	0 1 1
1 1 1 1 1 1	1 1 0
1 1 1 1 1 1 1	1 1 1 1 1 1 1 0
1 1 1 1 1 1 1 1	1 0 1 1 1 1 1 1

Set 2 Accept	Set 2 Reject
1 0	1
1 0 1 0	0
1 0 1 0 1 0	1 1
1 0 1 0 1 0 1 0	0 0
1 0 1 0 1 0 1 0 1 0 1 0	0 1
	1 0 1
	1 0 0
	1 0 0 1 0 1 0
	1 0 1 1 0
	1 1 0 1 0 1 0 1 0

Table 3. (cont.)

Set 3 Accept	Set 3 Reject
1	1 0
0	1 0 1
0 1	0 1 0
1 1	1 0 1 0
0 0	1 1 1 0
1 0 0	1 0 1 1
1 1 0	1 0 0 0 1
1 1 1	1 1 1 0 1 0
0 0 0	1 0 0 1 0 0 0
1 0 0 1 0 0	1 1 1 1 1 0 0 0
1 1 0 0 0 0 0 1 1 1 0 0 0 0 1	0 1 1 1 0 0 1 1 0 1
1 1 1 1 0 1 1 0 0 0 1 0 0 1 1 1 0 0	1 1 0 1 1 1 0 0 1 1 0

Set 4 Accept	Set 4 Reject
1	0 0 0
0	1 1 0 0 0
1 0	0 0 0 1
0 1	0 0 0 0 0 0 0 0
0 0	1 1 1 1 1 0 0 0 1 1
1 0 0 1 0 0	1 1 0 1 0 1 0 0 0 0 0 1 0 1 1 1
0 0 1 1 1 1 1 1 0 1 0 0	1 0 1 0 0 1 0 0 0 1
0 1 0 0 1 0 0 1 0 0	0 0 0 0
1 1 1 0 0	0 0 0 0 0
0 0 1 0	

Set 5 Accept	Set 5 Reject
1 1	0
0 0	1 1 1
1 0 0 1	0 1 0
0 1 0 1	0 0 0 0 0 0 0 0
1 0 1 0	1 0 0 0
1 0 0 0 1 1 1 1 0 1	0 1
1 0 0 1 1 0 0 0 0 1 1 1 1 0 1 0	1 0
1 1 1 1 1 1	1 1 1 0 0 1 0 1 0 0
0 0 0 0	0 1 0 1 1 1 1 1 1 1 1 0
	0 0 0 1
	0 1 1

Table 3. (cont.)

Set 6 Accept	Set 6 Reject
1 0	1
0 1	0
1 1 0 0	1 1
1 0 1 0 1 0	0 0
1 1 1	1 0 1
0 0 0 0 0 0	0 1 1
1 0 1 1 1	1 1 0 0 1
0 1 1 1 1 0 1 1 1 1	1 1 1 1
1 0 0 1 0 0 1 0 0	0 0 0 0 0 0 0
	0 1 0 1 1 1
	1 0 1 1 1 1 0 1 1 1 1 1
	1 0 0 1 0 0 1 0 0 1

Set 7 Accept	Set 7 Reject
1	1 0 1 0
0	0 0 1 1 0 0 1 1 0 0 0
1 0	0 1 0 1 0 1 0 1 0 1
0 1	1 0 1 1 0 1 0
1 1 1 1 1	1 0 1 0 1
0 0 0	0 1 0 1 0 0
0 0 1 1 0 0 1 1	1 0 1 0 0 1
0 1 0 1	1 0 0 1 0 0 1 1 0 1 0 1
0 0 0 0 1 0 0 0 0 1 1 1 1	
0 0 1 0 0	
0 1 1 1 1 1 0 1 1 1 1 1	
0 0	

Each training environment was simply defined by two sets of boolean strings, which are given in table 3. For uniformity, I ran all seven cases, as given, on a sequential cascaded network of a 1-input 4-output function network (with bias connections, making 8 weights for the context net to compute) and a 3-input 8-output context network with bias connections. The total of 32 context weights are essentially arranged as a 4 by 2 by 4 array. Only three of the outputs of the function net were fed back to the context network, while the fourth output unit was used as the accept bit. The standard back-propagation learning rate was set to 0.3 and the momentum to 0.7. All 32 weights were rest to random numbers between ± 0.5 for each run. Training was halted when all accept strings returned output bits above 0.8 and reject strings below 0.2.

Table 4. Minimal regular languages for the seven training sets.

Language #	Description
1	1*
2	(1 0)*
3	no odd zero strings after odd 1 strings
4	no 000's
5	pairwise, an even sum of 01's and 10's
6	number of 1's − number of 0's = 0 mod 3
7	0*1*0*1*

5.1. Results

Of Tomita's 7 cases, all but data-sets #2 and #6 converged without a problem in several hundred epochs. Case 2 would not converge, and kept treating negative case 110101010 as correct; I had to modify the training set (by adding reject strings 110 and 11010) in order to overcome this problem. Case 6 took several restarts and thousands of cycles to converge.

In the spirit of the machine learning community, I recently ran a series of experiments to make these results more empirical. Table 5 compares Tomita's stage one "number of mutations" to my "average number of epochs." Because back-propagation is sensitive to initial conditions (Kolen & Pollack, 1990), running each problem once does not give a good indication of its difficulty, and running it many times from different random starting weights can result in widely disparate timings. So I ran each problem 10 times, up to 1000 epochs, and average only those runs which separated the training sets (accepts above .6; rejects below .4). The column labeled "% Convergent" shows the percent of the 10 runs for each problem which separated the accept and reject strings within 100 cycles. Although it is difficult to compare results between completely different methods, taken together, the average epochs and the percent convergent numbers give a good idea of the difficulty of the Tomita data-sets for my learning architecture.

Table 5. Performance comparison between Tomita's Hill-climber and Pollack's model (Backprop).

Language	No. Mutations (Hill-Climber)	Avg. Epochs (Backprop)	% Convergent (Backprop)
1	98	54	100
2	134	787	20
3	2052	213	70
4	442	251	100
5	1768	637	80
6	277		0
7	206	595	50

5.2. Analysis

Tomita ran a brute-force enumeration to find the minimal automaton for each language, and verified that his hill-climber was able to find them. These are displayed in Figure 5. Unfortunately, I ran into some difficulty trying to figure out exactly which finite state automata (and regular languages) were being induced by my architecture.

For this reason, in figure 6, I present "prefixes" of the languages recognized and generated by the seven first-run networks. Each rectangle assigns a number, 0 (white) or 1 (black),

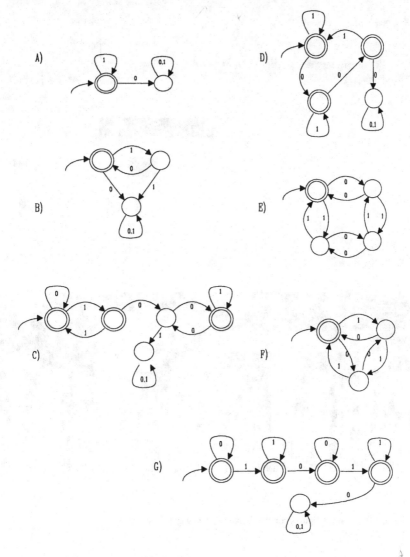

Figure 5. The minimal FSA's recognizing Tomita's 7 data sets.

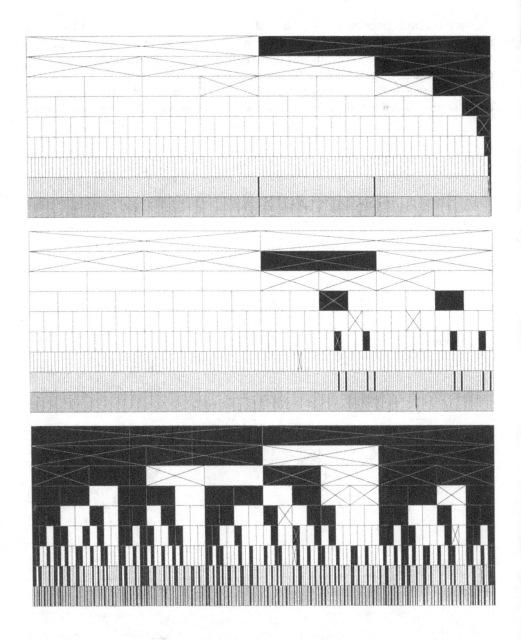

Figure 6. Recursive rectangle figures for the 7 induced languages. See text for detail.

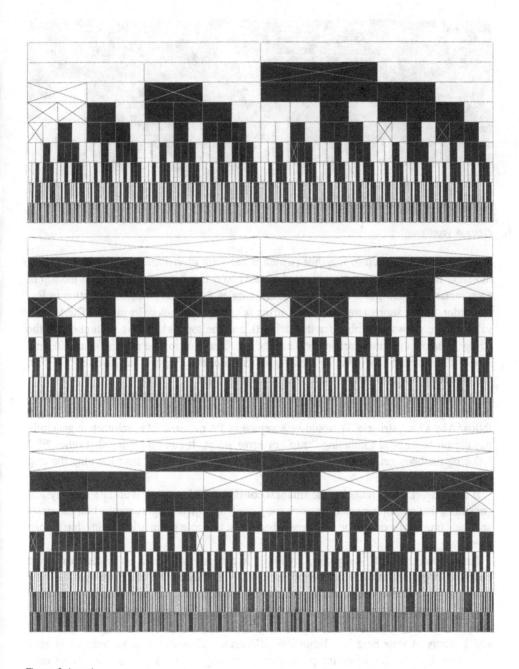

Figure 6. (cont.)

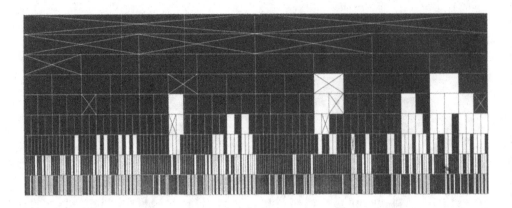

Figure 6. (cont.)

to all boolean strings up to length 9 (which is the limit of visibility), and thus indicates, in black, the strings which are accepted by the respective network. Starting at the top of each rectangle, each row r contains 2^r subrectangles for all the strings of length r in lexical order, so the subrectangle for each string is sitting right below its prefix. The top left subrectangle shows a number for the string 0, and the top right shows a number for the string 1. Below the subrectangle for 0 are the subrectangles for the strings 00 and 01, and so on. The training sets (table 4) are also indicated in these figures, as inverted "X's" in the subrectangles corresponding to the training strings.

Note that although the figures display some simple recursive patterns, none of the ideal minimal automata were induced by the architecture. Even for the first language 1*, a 0 followed by a long string of 1's would be accepted by the network. My architecture generally has the problem of not inducing "trap" or error states. It can be argued that other FSA inducing methods get around this problem by presupposing rather than learning the trap states.[3]

If the network is not inducing the smallest consistent FSA, what is it doing? The physical constraint that an implemented network use *finitely specified* weights means that the states and their transitions cannot be arbitrary—there must be some geometric relationship among them.

Based upon the studies of parity, my initial hypothesis was that a set of clusters would be found, organized in some geometric fashion: i.e., an embedding of a finite state machine into a finite dimensional geometry such that each token's transitions would correspond to a simple transformation of space. I wrote a program which examined the state space of these networks by recursively taking each unexplored state and combining it with both 0 and 1 inputs. A state here is a 3-dimensional vector, values of the three recurrently used output units. To remove floating-point noise, the program used a parameter ϵ and only counted states in each ϵ-cube once. Unfortunately, some of the machines seemed to grow drastically in size as ϵ was lowered. In particular, Figure 7 shows the log-log graph of the number of unique states versus ϵ for the machine resulting from training environment 7. Using the method of (Grassberger & Procaccia, 1983) this set was found to have a correlation dimension of 1.4—good evidence that it is "fractal."

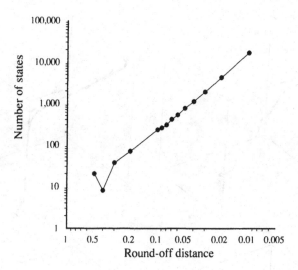

Figure 7. The number of states in the 7th machine grew dramatically as ϵ was lowered.

Because the states of the benchmark networks are "in a box" (Anderson et al., 1977) of low dimension, we can view these machines graphically to gain some understanding of how the state space is being arranged. Each 3-d vector is plotted as a point in the unit cube. Partial graphs of the state spaces for the first-run networks are shown in Figure 8. States were computed for all boolean strings up to and including length 10, so each figure contains 2048 points, often overlapping.

The images (a) and (d) are what I initially expected, clumps of points which closely map to states of equivalent FSAs. Images (b) and (e) have limit "ravines" which can each be considered states as well. However, the state spaces, (c), (f), and (g) of the dynamical recognizers for Tomita cases 3, 6, and 7, are interesting, because, theoretically, they are *infinite* state machines, where the states are not arbitrary or random, requiring an infinite table of transitions, but are constrained in a powerful way by mathematical principle.

In thinking about such a principle, consider systems in which extreme observed complexity emerges from algorithmic simplicity plus computational power. When I first saw some of the state space graphs (Figure 8), they reminded me of Barnsley's Iterated Functional Systems (Barnsley, 1988), where a compactly coded set of affine transformations is used to *iteratively* construct displays of fractals, previously described *recursively* using line-segment rewrite rules (Mandelbrot, 1982). The calculation is simply the repetitive transformation (and plotting) of a state vector by a sequence of randomly chosen affine transformations. In the infinite limit of this process, fractal "attractors" emerge (e.g., the widely reproduced fern).[4]

By eliminating the sigmoid, commuting the y_j and z_k terms in Eq. 1:

$$z_i(t) = \sum_k \left(\sum_j w_{ijk}\, y_j(t) \right) x_k(t-1)$$

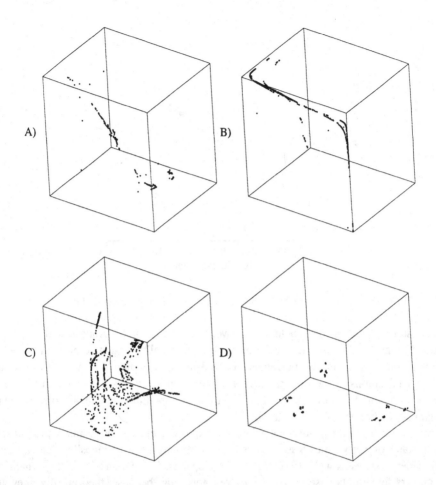

Figure 8. Images of the state spaces for the 7 Tomita training environments. Each box contains 2048 points, the states corresponding to all boolean strings up to length 10.

and treating the y_j's as an infinite random sequence of binary unit vectors (1-in-j codes), the forward pass calculation for my network can be seen as the same process used in an Iterated Function System (IFS). Thus, my figures of state-spaces, which emerge from the projection of Σ^* into Z, are fractal attractors, as defined by Barnsley.

6. Related work

The architecture and learning paradigm I used is also being studied by Lee Giles and colleagues, and is closely related to the work of Elman and Servan-Schreiber et al on Simple Recurrent Networks. Both architectures rely on extending Jordan's recurrent networks in a direction which separates visible output states from hidden recurrent states, without making

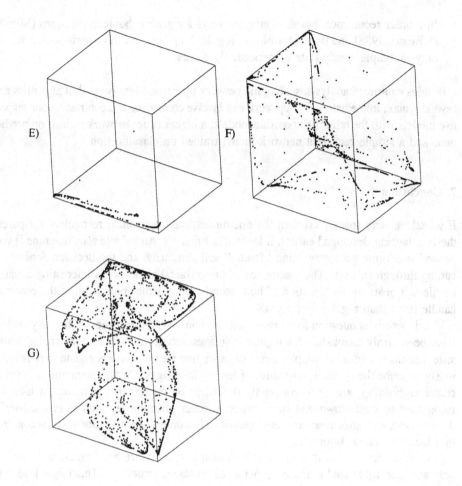

Figure 8. (cont.)

the unstable "back-propagation through time" assumption. Besides our choice of language data to model, the two main differences are that:

(1) They use a "predictive" paradigm, where error feedback is provided at every time step in the computation, and I used a "classification" paradigm, feeding back only at the end of the given examples. Certainly, the predictive paradigm is more psychologically plausible as a model of positive only presentation (c.f., Culicover & Wexler, pp. 63–65), but the Tomita learning environments are much more impoverished.

I have no commitment to negative information; all that is required is some desired output which discriminates among the input strings in a generalizable way. Positive versus negative evidence is merely the simplest way (with 1 bit) to provide this discrimination.

(2) They use a single layer (first order) recurrence between states, whereas I use a higher order (quadratic) recurrence. The multiplicative connections are what enable my model to have "fractal" dynamics equivalent in the limit to an IFS, and it may be that the

first-order recurrence, besides being too weak for general boolean functions (Minsky & Papert, 1988) and thus for arbitrary regular languages (such as parity), also results only in simple steady-state or periodic dynamics.

Besides continued analysis, scaling the network up beyond binary symbol alphabets and beyond syntax, immediate follow-up work will involve comparing and contrasting our respective models with the other two possible models, a higher order network trained on prediction, and a simple recurrent network model trained on classification.

7. Conclusion

If we take a state space picture of the one-dimensional dynamical recognizer for parenthesis balancing developed earlier, it looks like Figure 9. An infinite state machine is embedded in a finite geometry using "fractal" self-similarity, and the decision function is cutting through this set. The emergence of these fractal attractors is interesting because I believe it bears on the question of how neural-like systems could achieve the power to handle more than regular languages.

This is a serious question for connectionism to answer, because since Chomsky (1956), it has been firmly established that regular languages, recognizable by Markov chains, finite-state machines, and other simple iterative/associative means, are inadequate to parsimoniously describe the syntactic structures of natural languages. Certain phenomena, such as center embedding, are more compactly described by context-free grammars which are recognized by Push-down Automata, whereas other phenomena, such as crossed-serial dependencies and agreement, are better described by context-sensitive grammars, recognized by Linear Bounded Automata.

On the one hand, it is quite clear that human languages are not formal, and thus are only analogically related to these mathematical syntactic structures. This might lead connectionists to erroneously claim that recursive computational power is not of the "essence of human computation."[5] It is also quite clear that without understanding these complexity issues, connectionists can stumble again and again into the trap of making strong claims for their models, easy to attack for not offering an adequate replacement for established theory. (Fodor & Pylyshyn, 1988; Pinker & Prince, 1988). But it is only because of "long-term lack of competition" that descriptive theories involving rules and representations can be defended as explanatory theories. Here is an alternative hypothesis for complex syntactic structure:

> The state-space limit of a dynamical recognizer, as $\Sigma^* \to \Sigma^\infty$, is an Attractor, which is cut by a threshold (or similar decision) function. The complexity of the generated language is regular if the cut falls between disjoint limit points or cycles, context-free if it cuts a "self-similar" (recursive) region, and context-sensitive if it cuts a "chaotic" (pseudo-random) region.

There is certainly substantial need for work on the theoretical front to more thoroughly formalize and prove or disprove the six main theorems implied by my hypothesis. I do

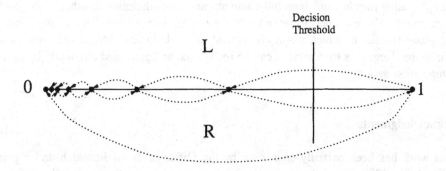

Figure 9. Slicing through the "fractal" state space of the balanced parenthesis dynamical recognizer.

not expect the full range of context-free or context sensitive systems to be covered by conventional quasi-linear processing constraints, and the question remains wide open as to whether the syntactic systems which can be described by neural dynamical recognizers have any convergence with the needs of natural language systems.

Because information processing provides the "essence" of complex forms of cognition, like language, it is important to understand the relationship between complex emergent behaviors of dynamical systems (including neural systems) and traditional notions of computational complexity, including the Chomsky hierarchy as well as algorithmic information theory (Chaitin, 1966), but the study of this relationship is still in its infancy.

In (Pollack, 1987b), I constructed a Turing Machine out of connectionist parts, and essentially showed that rational values, constants, precise thresholds, and multiplicative connections (all used in the sequential cascaded network architecture) were sufficient primitives for computationally universal recurrent neural networks.

Cellular Automata, which we might view as a kind of low-density, synchronous, uniform, digital restriction of neural networks, have been studied as dynamical systems (Wolfram, 1984) and proven to be as powerful as universal Turing Machines, e.g., (Lindgren & Nordahl, 1990). Furthermore, (Moore, 1990) has shown that there are simple mathematical models for dynamical systems which are also universal, and it follows directly that determination of the behavior of such dynamical systems in the limit is undecidable and unpredictable, even with precise initial conditions. In stronger terms, the theoretical foundations of computer and information science may be in accord with the lack of predictability in the universe.

Finally, Crutchfield and Young (1989) have studied the computational complexity of dynamical systems reaching the onset of chaos via period-doubling. They have shown that these systems are not regular, but are finitely described by Indexed Context-Free Grammars. It may, of course, be just a coincidence that several modern computational linguistic grammatical theories also fall in this class (Joshi, 1985; et al., 1989; Pollard, 1984).

In conclusion, I have merely illuminated the possibility of the existence of a naturalistic alternative to explicit recursive rules as a description for the complexity of language. Such a mathematical description would be compact, "parsimonious" in fact, since the infinite state machine does not require infinite description, but only a finitely described set of weights. It was shown to be feasible to learn this type of description from a finite set of examples using pseudo-continuous hill-climbing parameter-adaptation (in other words, back-propagation). However, performance in the limit appears to jump in discrete steps, inductive phase transitions which might correspond to psychological "stages" of acquisition. Finally, the languages so described can be recognized and generated efficiently by neural computation systems.

Acknowledgments

This work has been partially sponsored by the Office of Naval Research under grant N00014-89-J-1200. Thanks to the numerous colleagues who have discussed and/or criticized various aspects of this work or my presentation of it, including: T. Bylander, B. Chandrasekaran, J. Crutchfield, L. Giles, E. Gurari, S. Hanson, R. Kasper, J. Kolen, W. Ogden, T. Patten, R. Port, K. Supowit, P. Smolensky, D.S. Touretzky, and A. Zwicky.

Notes

1. To turn a recognizer into a generator, simply enumerate the strings in Σ^* and filter out those the recognizer rejects.
2. For the simple low dimensional dynamical systems usually studied, the "knob" or control parameter for such a bifurcation diagram is a scalar variable; here the control parameter is the entire 32-D vector of weights in the network, and back-propagation turns the knob.
3. Tomita assumed a trap state which did not mutate, and (Servan-Schreiber et al., 1989) compared the incoming token to a thresholded set of token predictions, trapping if the token was not predicted.
4. Barnsley's use of the term "attractor" is different than the conventional use of the term given in the introduction, yet is technically correct in that it refers to the "limit" of an iterative process. It can be thought of as what happens when you randomly drop an infinite number of microscopic iron filing "points" onto a piece of paper with a magnetic field lying under it; each point will land "on" the underlying attractor.
5. (Rumelhart & McClelland, 1986), p. 119.

References

Anderson, J.A., Silverstein, J.W., Ritz, S.A. & Jones, R.S. (1977). Distinctive features, categorical perception, and probability learning: Some applications of a neural model. *Psychological Review, 84*, 413–451.

Angluin, D. (1978). On the complexity of minimum inference of regular sets. *Information and Control, 39*, 337–350.

Angluin, D. & Smith, C.H. (1983). Inductive inference: Theory and methods. *Computing Surveys, 15*, 237–269.

Barnsley, M.F. (1988). *Fractals everywhere*. San Diego: Academic Press.

Berwick, R. (1985). *The acquisition of syntactic knowledge*. Cambridge: MIT Press.

Chaitin, G.J. (1966). On the length of programs for computing finite binary sequences. *Journal of the AM, 13*, 547–569.

Chomsky, N. (1956). Three models for the description of language. *IRE Transactions on Information Theory, IT-2*, 113–124.

Chomsky, N. (1965). *Aspects of the theory of syntax*. Cambridge, MA: MIT Press.

Crutchfield, J.P., Farmer, J.D., Packard, N, H. & Shaw, R.S. (1986). Chaos. *Scientific American, 255*, 46–57.

Crutchfield, J.P. & Young, K. (1989). Computation at the onset of chaos. In W. Zurek, (Ed.), *Complexity, entropy and the physics of Information*. Reading, MA: Addison-Wesley.

Derrida, B. & Meir, R. (1988). Chaotic behavior of a layered neural network. *Phys. Rev. A, 38.*

Devaney, R.L. (1987) *An introduction to chaotic dynamical systems.* Reading, MA: Addison-Wesley.

Elman, J.L. (1990). Finding structure in time. *Cognitive Science, 14,* 179–212.

Feldman, J.A. (1972). Some decidability results in grammatical inference. *Information & Control, 20,* 244–462.

Fodor, J. & Pylyshyn, A. (1988). Connectionism and cognitive architecture: A critical analysis. *Cognition, 28,* 3–71.

Giles, C.L., Sun, G.Z., Chen, H.H., Lee, Y.C. & Chen, D. (1990). Higher order recurrent networks and grammatical inference. In D.S. Touretzky, (Ed.), *Advances in neural information processing systems.* Los Gatos, CA: Morgan Kaufmann.

Gleick, J. (1987). *Chaos: Making a new science.* New York: Viking.

Gold, E.M. (1967). Language identification in the limit. *Information & Control, 10,* 447–474.

Gold, E.M. (1978). Complexity of automaton identification from given data. *Information and Control, 37,* 302–320.

Grassberger, P. & Procaccia, I. (1983). Measuring the strangeness of strange attractors. *Physica, 9D,* 189–208.

Grebogi, C., Ott, E. & Yorke, J.A. (1987). Chaos, strange attractors, and fractal basin boundaries in nonlinear dynamics. *Science, 238,* 632–638.

Hendin, O., Horn, D. & Usher, M. (1991). Chaotic behavior of a neural network with dynamical thresholds. *Int. Journal of Neural Systems, to appear.*

Hopfield, J.J. (1982). Neural networks and physical systems with emergent collective computational abilities. *Proceedings of the National Academy of Sciences USA, 79,* 2554–2558.

Hornik, K., Stinchcombe, M. & White, H. (1990). Multi-layer feedforward networks are universal approximators. *Neural networks, 3.*

Huberman, B.A. & Hogg, T. (1987). Phase transitions in artificial intelligence systems. *Artificial Intelligence, 33,* 155–172.

Joshi, A.K. (1985). Tree adjoining grammars: How much context-sensitivity is required to provide reasonable structural descriptions? In D.R. Dowty, L. Karttunen & A.M. Zwicky, (Eds.), *Natural language parsing.* Cambridge, Cambridge University Press.

Joshi, A.K., Vijay-shanker, K. & Weir, D.J. (1989). Convergence of mildly context-sensitive grammar formalism. In T. Wasow & P. Sells, (Eds.), *The processing of linguistic structure.* Cambridge: MIT Press.

Kolen, J.F. & Pollack, J.B. (1990). Back-propagation is sensitive to initial conditions. *Complex Systems, 4,* 269–280.

Kurten, K.E. (1987). Phase transitions in quasirandom neural networks. In *Institute of Electrical and Electronics Engineers First International Conference on Neural Networks.* San Diego, II-197-20.

Lapedes, A.S. & Farber, R.M. (1988). *How neural nets work* (LAUR-88-418): Los Alamos, NM.

Lieberman, P. (1984). *The biology and evolution of language.* Cambridge: Harvard University Press.

Lindgren, K. & Nordahl, M.G. (1990). Universal computation in simple one-dimensional cellular automata. *Complex Systems, 4,* 299–318.

Lippman, R.P. (1987). An introduction to computing with neural networks. *Institute of Electrical and Electronics Engineers ASSP Magazine, April,* 4–22.

MacLennan, B.J. (1989). *Continuous computation* (CS-89-83). Knoxville, TN: University of Tennessee, Computer Science Dept.

MacWhinney, B. (1987). *Mechanisms of language acquisition.* Hillsdale: Lawrence Erlbaum Associates.

Mandelbrot, B. (1982). *The fractal geometry of nature.* San Francisco: Freeman.

McCulloch, W.S. & Pitts, W. (1943). A logical calculus of the ideas immanent in nervous activity. *Bulletin of Mathematical Biophysics, 5,* 115–133.

Mealy, G.H. (1955). A method for synthesizing sequential circuits. *Bell System Technical Journal, 43,* 1045–1079.

Metcalfe, J. & Wiebe, D. (1987). Intuition in insight and noninsight problem solving. *Memory and Cognition, 15,* 238–246.

Minsky, M. (1972). *Computation: Finite and infinite machines.* Cambridge, MA: MIT Press.

Minsky, M. & Poper, S. (1988). *Perceptrons.* Cambridge, MA: MIT Press.

Moore, C. (1990). Unpredictability and undecidability in dynamical systems. *Physical Review Letters, 62,* 2354–2357.

Mozer, M. (1988). *A focused back-propagation algorithm for temporal pattern recognition* (CRG-Technical Report-88-3). University of Toronto.

Pearlmutter, B.A. (1989). Learning state space trajectories in recurrent neural networks. *Neural Computation, 1,* 263–269.

Pineda, F.J. (1987). Generalization of back-propagation to recurrent neural networks. *Physical Review Letters, 59,* 2229–2232.

Pinker, S. (1984). *Language learnability and language development*. Cambridge: Harvard University Press.

Pinker, S. & Prince, A. (1988). On language and connectionism: Analysis of a parallel distributed processing model of language inquisition. *Cognition, 28*, 73–193.

Pinker, S. & Bloom, P. (1990). Natural language and natural selection. *Brain and Behavioral Sciences, 12*, 707–784.

Plunkett, K. & Marchman, V. (1989). *Pattern association in a back-propagation network: Implications for child language acquisition* (Technical Report 8902). San Diego: UCSD Center for Research in Language.

Pollack, J.B. (1987). Cascaded back propagation on dynamic connectionist networks. *Proceedings of the Ninth Conference of the Cognitive Science Society* (pp. 391–404). Seattle, WA.

Pollack, J.B. (1987). *On connectionist models of natural language processing*. Ph.D. Thesis, Computer Science Department, University of Illinois, Urbana, IL. (Available as MCCS-87-100, Computing Research Laboratory, Las Cruces, NM)

Pollack, J.B. (1989). Implications of recursive distributed representations. In D.S. Touretzky, (Ed.), *Advances in neural information processing systems*. Los Gatos, CA: Morgan Kaufmann.

Pollack, J.B. (1990). Recursive distributed representation. *Artificial Intelligence, 46*, 77–105.

Pollard, C. (1984). *Generalized context-free grammars, head grammars and natural language*. Doctoral Dissertation, Dept. of Linguistics, Stanford University, Palo Alto, CA.

Rivest, R.L. & Schapire, R.E. (1987). A new approach to unsupervised learning in deterministic environments. *Proceedings of the Fourth International Workshop on Machine Learning* (pp. 364–475). Irvine, CA.

Rumelhart, D.E. & McClelland, J.L. (1986). PDP models and general issues in cognitive science. In D.E. Rumelhart, J.L. McClelland & the PDP Research Group, (Eds.), *Parallel distributed processing: Experiments in the microstructure of cognition*, Vol. 1. Cambridge: MIT Press.

Rumelhart, D.E., Hinton, G. & Williams, R. (1986). Learning internal representations through error propagation. In D.E. Rumelhart, J.L. McClelland & the PDP Research Group, (Eds.), *Parallel distributed processing: Experiments in the microstructure of cognition*, Vol. 1. Cambridge: MIT Press.

Servan-Schreiber, D., Cleeremans, A. & McClelland, J.L. (1989). Encoding sequential structure in simple recurrent networks. In D.S. Touretzky, (Ed.), *Advances in neural information processing systems*. Los Gatos, CA: Morgan Kaufmann.

Skarda, C.A. & Freeman, W.J. (1987). How brains make chaos. *Brain & Behavioral Science, 10.*

Smolensky, P. (1986). Information processing in dynamical systems: Foundations of harmony theory. In D.E. Rumelhart, J.L. McClelland & the PDP Research Group, (Eds.), *Parallel distributed processing: Experiments in the microstructure of cognition*, Vol. 1. Cambridge: MIT Press.

Tomita, M. (1982). Dynamic construction of finite-state automata from examples using hill-climbing. *Proceedings of the Fourth Annual Cognitive Science Conference* (pp. 105–108). Ann Arbor, MI.

Touretzky, D.S. & Geva, S. (1987). A distributed connectionist representation for concept structures. *Proceedings of the Ninth Annual Conference of the Cognitive Science Society* (pp. 155–164). Seattle, WA.

van der Maas, H., Verschure, P. & Molenaar, P. (1990). A note on chaotic behavior in simple neural networks. *Neural Networks, 3*, 119–122.

Wexler, K. & Culicover, P.W. (1980). *Formal principles of language acquisition*. Cambridge: MIT Press.

Wolfram, S. (1984). Universality and complexity in cellular automata. *Physica, 10D*, 1–35.

INDEX